AF710457

WITHDRAWN

Engineering Analysis of Vehicular Accidents

Randall K. Noon

CRC

CRC Press
Boca Raton Ann Arbor London Tokyo

HV
8079.55
.N66
1994

Library of Congress Cataloging-in-Publication Data

Noon, Randall.
Engineering analysis of vehicular accidents / Randall K. Noon.
p. cm.
Includes bibliographical references and index.
ISBN 0-8493-8104-5
1. Traffic accident investigation. 2. Engineering mathematics.
I. Title.
HV8079.55.N66 1994 93-48723
363.12′565—dc20 CIP

This book contains information obtained from authentic and highly regarded sources. Reprinted material is quoted with permission, and sources are indicated. A wide variety of references are listed. Reasonable efforts have been made to publish reliable data and information, but the author and the publisher cannot assume responsibility for the validity of all materials or for the consequences of their use.

Neither this book nor any part may be reproduced or transmitted in any form or by any means, electronic or mechanical, including photocopying, microfilming, and recording, or by any information storage or retrieval system, without prior permission in writing from the publisher.

CRC Press, Inc.'s consent does not extend to copying for general distribution, for promotion, for creating new works, or for resale. Specific permission must be obtained in writing from CRC Press for such copying.

Direct all inquiries to CRC Press, Inc., 2000 Corporate Blvd., N.W., Boca Raton, FL 33431.

© 1994 by CRC Press, Inc.

No claim to original U.S. Government works
International Standard Book Number 0-8493-8104-5
Printed in the United States of America 1 2 3 4 5 6 7 8 9 0
Printed on acid-free paper

Foreword

This text is about applying engineering skills, knowledge, and judgment to analyze vehicular accidents. It assumes that the reader is a student of engineering or science who has had course work in statics, dynamics, calculus, vector algebra, and perhaps basic strength of materials. In that regard, it is likely that no new fundamental scientific concepts will be introduced to the reader, just new applications.

The general approach used in the technical chapters is simple. First, the general principles are reviewed. Second, the relevant equations are derived from the general principles. Third, the significance of the derived equations is discussed or demonstrated.

Three chapters are notably not filled with copious equations and derivations — Chapters 1, 6, and 12. Regrettably, many students of engineering and science tend to underrate such chapters, assuming that if there are no equations, the material must be of small importance. However, I urge the reader to consider carefully the ideas and topics contained in them.

Accident analysis is not like a classroom problem, where the equations are selected and solved, and the matter is considered finished. Many novices in this field of work are incredulous when a jury decides for the "other side," especially when they have prepared a page full of calculations that "obviously" proves that the jury's decision is wrong. Engineering skill is a must, but there is no substitute for skillful practice.

Chapter 9 discusses at some length the application of the conservation of momentum to solving impact problems. This method is most accommodating to police accident reports. A well prepared police accident report usually notes skid marks, final vehicle positions, initial travel directions, and the point of impact. This constitutes the usual data needed to solve momentum equations. Chapter 9 also introduces the use of the coefficient of restitution to assess fixed barrier impacts and other types of impacts. While the coefficient of restitution is generally underutilized in accident reconstruction, it is a powerful solution tool when skillfully applied.

Chapter 10 is notable in showing a simplified energy analysis method. More traditional energy analysis methods tend to use a first-order or second-order vehicle stiffness function to derive a ΔV versus crush deformation relation. Such a polynomial stiffness function is necessary since the plot of ΔV versus crush deformation is a curve. By using, instead, a kinetic energy versus crush deformation relation, the stiffness function becomes a constant and the analysis is greatly simplified.

At the end of some of the chapters are example problems, based largely on actual problems regularly encountered in practice. The example problems should not be considered separate from the chapter, but as a continuation of it. Many of the example problems contain tips or concepts not discussed in the chapter. To not consider the sample problems is the same as not reading a portion of the chapter.

It has been my intention to prepare a working, instructional text rather than a strictly academic one. Therefore, I have used footnotes sparingly. The traditional practice of inserting copious footnotes in the text to demonstrate the author's erudition is generally distracting to the reader, and makes the task of absorbing new material even more difficult. However, no work such as this is done in a vacuum, and I would be remiss if I did not acknowledge the body of work upon which I have drawn. For this reason, Appendix I contains an ample list of references which the reader can use to verify and augment concepts and information introduced in the text.

At the close of a preface, it is customary to thank those persons who have aided the author in preparing the text. In that regard, my friends, cohorts, and relatives all know how sincerely indebted I am to them, especially my wife, Leslie. I am sure they are all glad they won't be asked to proofread a chapter or two for a while. However, I also wish to sincerely thank Presidents Kennedy and Johnson, the United States Congress, and the American taxpayers. These good-hearted people provided the loans, grants, and scholarships that allowed this ordinary baker's son a chance to go to college. On behalf of myself, my generation, and the children of my generation, I sincerely thank you.

R.K.N.
September 1, 1993

About the Author

Randall Noon is currently the Director of Engineering Services at the firm of Dressler Consulting Engineers, Inc. in Overland Park, Kansas. In this capacity Randy directs the engineering activities of the firm related to investigative engineering, which includes vehicular accident reconstruction. Randy holds a B.S.M.E. from the University of Missouri-Columbia, an M.S.M.E. from the University of Missouri--Rolla, and a Ph.D. in engineering from Pacific Western University. Randy is professionally licensed in both the U.S. and Canada, and is also the author of "*Introduction to Forensic Engineering*", also published by CRC Press in 1992.

Randy lives in the countryside near Topeka, Kansas, with his charming wife Leslie, his beautiful daughter Molly, two dogs, two cats, and assorted denizens of the woods. His wife Leslie warns all of you not to ask him at a party about the significance of Edison on twentieth century life.

Warranties

No warranty is implied or expressed by the publication of this book. While every effort has been made to ensure accuracy and correctness, the publication is provided for instructional purposes only. Appropriate use of the materials presented is at the discretion of the reader, who assumes full responsibility.

Table of Contents

CHAPTER 1

Introduction

A. General

Vehicular accident reconstruction is the application of scientific knowledge and methodology to determine what happened in an accident. Often, the specific factors to be determined include the speed of the involved vehicles upon impact, the speed of the vehicles prior to braking, the initial direction vectors of the vehicles, the field of vision of the drivers, the condition of the vehicles before impact, the location of impact, the potential for evading the accident, the time durations of accident events, and so on.

Generally, to maintain a verifiable basis for analysis, there should be primary reliance upon the actual physical evidence. Important items include tire and skid marks, a plan of the accident area, the location and depth of impact on the vehicles, the condition of the vehicles, the type and location of impact debris, paint "tags" and other vehicle-to-vehicle impact marks, road conditions, weather conditions, and the condition of the drivers. Other information that could bear upon the causation of the accident includes vehicle model performance parameters, roadway specifications, type and placement of traffic control signs, and date and time of the accident.

While eyewitness accounts of the accident are important sources of information, they must be carefully scrutinized. Sometimes eyewitnesses to the same accident disagree on the most fundamental aspects of the accident. They will intertwine their own conclusions about the accident into their narrative of what happened. Also, an eyewitness may not be impartial: the eyewitness may be the wife, boyfriend, cousin, close friend, or enemy of one of the parties involved in the accident. A proper scientific accident reconstruction can often provide information to discriminate between who is telling the truth, and who is not.

The scientific disciplines often applied in accident reconstruction include kinematics, dynamics, mechanics of materials, machine design, ergonometrics, and traffic engineering. These disciplines are normally a part of traditional mechanical and civil engineering curricula, although the specific applications are different from those shown in the classroom.

For example, a typical undergraduate civil engineering course in mechanics of materials concentrates on the factors that contribute to the initiation of failure. Generally, the elastic properties of the material are the most studied. In an accident reconstruction, however, the structure being studied, the vehicle, often has already failed in a material sense; it has been crushed by impact. The more important material properties in this case are the plastic properties of the material.

B. Role in the Legal System

From time to time a person who does this type of engineering analysis, a forensic engineer or accident reconstructionist, is called upon to testify in deposition or court about the specifics of an accident reconstruction. Normally the testimony consists of answers to questions posed by an attorney for one of the parties involved. The attorney will often be interested in the following:

- the engineer's qualifications to perform accident reconstruction analysis,
- the basic facts and assumptions that were relied upon to do the analysis,
- the reasonableness of the engineer's conclusions, and
- plausible alternative explanations for the accident that may not have been considered by the engineer (which often will be the attorney's client's version of the accident).

By virtue of the appropriate education and experience, a person may be qualified by the court as an "expert witness". In some states, such an expert witness is the only person allowed to render an opinion to the court during proceedings. Because the U.S. legal system is adversarial in nature, each attorney will try to elicit from the expert witness testimony that will either benefit his own client, or disparage his adversary's client.

In such a role, despite the fact that one of the attorneys is paying the expert's fee, the expert witness has an obligation to the court to be as objective as possible, and to refrain from being an advocate for either side. The best rule to follow is to be honest and professional both in preparing the original analysis of the accident, and in testifying. Before giving testimony, however, the expert witness has an obligation to fully discuss with his or her client both the favorable and the unfavorable aspects of the accident analysis.

Sometimes the engineer involved in preparing an accident analysis is requested to review a report or analysis of the accident that was prepared by an expert witness for the "other side." This should also be done honestly and professionally. Petty qualifications, "one-upmanship", personal attacks, and unfounded criticisms are unproductive and can prove embarrassing to the person who engages in them. When preparing a criticism of someone else's work, consider what it would sound like when read to a jury in open court.

Honest disagreements between two qualified experts do happen. When such disagreements occur, the focus of criticism should be the theoretical or factual basis for the differences.

C. The Scientific Method

Roger Bacon, a 13th-century English Franciscan monk, is often credited with defining the modern scientific method. He believed that scientific knowledge should be obtained by close observation and experimentation. He experimented with gunpowder, lodestones, and optics to mention just a few items, and was dubbed "Doctor Admirabilis" because of his extensive knowledge.

For his efforts to put knowledge on a verifiable basis, the curious friar was accused of necromancy, heresy, and black magic by the chiefs of his monastic order. He was confined to a monastery in Paris for ten years so that he could be watched. His efforts to persuade Pope Clement IV to allow experimental science to be taught at the university failed. After Pope Clement IV died, Bacon was imprisoned for another ten years. Pope Nicholas III forbade the reading of Bacon's papers and books, which were generally banned already. Some scholars believe that Roger Bacon was the original inspiration for the Doctor Faustus legend.

Why all the fuss over a friar who wants to play in the laboratory? It is because he threatened the "correctness" of theories promulgated by the Church. At the time of Roger Bacon, the doctrine of *apriorism* was the accepted basis for all inquiry. Apriorism is the belief that the underlying causes for all observed effects are already known, or at least can be deduced from some first principles. Under apriorism, if a person's observations conflict with the accepted theory, then the observations must be imperfect, or are the product of the Devil tempting a foolish mortal.

For example, in Bacon's time it was assumed *a priori* that the sun revolved about the earth and all celestial motions were based on perfect circles, and all other theories concerning the universe had to encompass these assumptions. This led, of course, to many inaccuracies and difficulties in accounting for the motions of the planets. Why were these things assumed to be correct in the first place? It was because they were deemed "reasonable" and compatible with accepted dogma. In fact, various Bible verses were cited to "prove" these assumptions. Thus, to cast doubt upon the assumption of the earth being at the center of the universe was to cast doubt upon the Church itself.

Fortunately, we no longer burn people at the stake for suggesting that the sun is the center of our solar system, or that planets have elliptical orbits. The modern scientific method does not accept the *a priori* method of inquiry, and works as follows: First, careful and detailed observations of actual events are made. Then, based upon the observations, a *working hypothesis* is formulated to explain the observations. Experiments or additional observations are then made to test the predictive ability of the working hypothesis.

As more observations are collected and studied, it may be necessary to modify, amplify or even discard the original hypothesis in favor of a new one which can better account for all the observations and data. A hypothesis is not considered valid unless it accounts for all the relevant observations and data, unless of course the data or observations are proved to be inaccurate.

D. Application to Accident Reconstruction

The scientific method is directly applicable to vehicular accident reconstruction. Conclusions about an accident should be consistent with the whole body of available physical evidence and verifiable observations, and should be consistent with proven scientific principles.

The doctrine of apriorism, unfortunately, is often applied to accident reconstruction work by a number of poor practitioners. This occurs when the investigator has decided ahead of time what happened, and simply looks for a few facts to support his or her initial assumptions. It is worth remembering that people who believe the earth is flat can also point to a small number of observations to prove their claim.

E. Magnitude of Accident Problem

In the United States, approximately 50,000 people die each year in automobile accidents. The following table lists the yearly deaths from automobile accidents from a recent six-year period.

Table 1. U.S. Motor Vehicle Deaths

year	number of deaths
1980	53,174
1981	51,385
1982	45,779
1983	44,452
1984	46,263
1985	45,600
1986	47,900

As a comparison, there were 47,382 U.S. battle deaths reported for the entire Viet Nam War. In World War I, there were a total of 53,402 U.S. battle deaths. During the five-year period of World War II, there were 291,557 U.S. battle deaths: an average of 58,311 per year. As the numbers show, the annual carnage on our country's highways has about the same magnitude as the carnage our country has experienced in war.

In 1986, the total death rate for the entire U.S. was 8.8 per 1,000 people. Deaths resulting only from motor vehicle accidents occurred at a rate of 0.26 per 1,000 people. In short, some 3% of all the deaths in the U.S. result from motor vehicle accidents, and this makes vehicle accidents one of our major life hazards.

Of course, the above death rates do not include the number of non-fatal injuries associated with motor vehicle accidents. As a rule of thumb, for every motor vehicle death about forty people are injured.

The following table lists the factors involved in motor vehicle accidents in the U.S. in 1986. These figures give a rough guide as to the general distribution of factors causing accidents. As noted, improper driving was a factor in over half of the accidents.

Figure 1. One-car accident caused by other vehicle whose driver was oblivious to the situation. Phone company truck (2nd on right) avoided a collision but hit guard rail.

Table 2. Accident Factors in 1986

Category	Injury accident	Fatal accident
speeding	22.4%	25.5%
right of way	21.9	10.8
driving left of center	2.3	8.4
improper passing	1.3	1.4
other forms of improper driving	15.3	9.8
total improper driving	63.2%	55.9%
other causes	36.8%	44.1%

From the above table, it is seen why the speed, position, direction and so on of the vehicles involved in an accident are major concerns in reconstructing accidents. Excessive speed alone is a factor in roughly a quarter of all motor vehicle accidents.

F. Setting Limits

While it may not always be possible to have available all the necessary information to determine every factor in an accident, it is not always necessary to do so. In fact, it is often the case that sufficient information about the accident can be determined by setting limits as to what is possible, and what is not.

As an example, the question might be posed as to whether or not a driver exceeded the posted speed limit of 40 mph. Perhaps there is insufficient information to determine the exact speed of the vehicle, but there is ample information to indicate that the vehicle was travelling at least 45 mph or more.

Similarly, one eyewitness may contend that vehicle "A" went from a dead stop to location "1" in a certain time. However, a review of the published acceleration parameters of the car may demonstrate that it is not capable of such an action; this, of course, would cast doubt on the veracity of the eyewitness' testimony.

In many situations, applying scientific principles or facts can provide upper or lower limits to the possible actions of the involved vehicles. This in itself may be sufficient for the case. It may provide a basis to exclude certain contentions, or affirm the possibility of others.

Figure 2. Fire Damaged Pickup. Less than 1% of accidents involve fires; most fires are caused by fuel leaks in the engine compartment.

G. Available Information

It is rare that an engineer has the opportunity to examine the accident scene before it is cleaned up. In fact, it is typical for an engineer to be assigned to analyze an accident days, weeks or even months after the actual event.

In general, the following information is typically available to assist in analyzing an accident:

- police accident report,
- photographs of the accident scene and accident vehicles taken by the police or insurance adjusters,
- the accident vehicles which have been towed to a salvage lot, and
- statements of the involved parties and witnesses.

A good police accident report is generally fundamental to a good accident reconstruction analysis. In addition to providing the basic information of "when, where, and who," a good police report will:

- diagram the position of the vehicles as found after the accident,
- diagram the tire marks, impact marks, impact debris, and other items found at the scene,
- contain statements made by the involved parties or witnesses directly after the accident.

After an accident, the vehicles are normally impounded for a time by the local authorities in a salvage yard. This may provide an opportunity to make measurements, perform tests, photograph, and otherwise document the condition of the vehicles. It is a good idea to corroborate marks or debris noted by the police at the accident scene in the accident report to damages observed and measured on the vehicles.

Whenever possible, it is best to view first-hand the vehicles involved in the accident. This is true even if the initial analysis of the police report data indicates that the causes of the accident are cut and dry.

Even the contents of a vehicle can provide insight into the cause of an accident. For example, finding a spilt soft drink all over the dashboard in front of the driver may suggest that the driver was taking a drink at the time of impact, and may have not seen the danger. Finding an open road map may suggest that the driver was in an unfamiliar area and was lost.

In getting the accident vehicles from the accident scene to the salvage yard, sometimes additional damages or modifications occur that are unrelated to the accident itself. For example, doors may have been cut off to extricate victims, flat tires may have been fixed or wheels replaced to facilitate towing. The vehicles may have been roughly handled, dragged, or dropped in order to be placed in rows or particular parking slots in the salvage yard. Such post-accident changes to the vehicles have to be noted and segregated from the analysis of the actual accident damages.

Some accident reconstructions are have to be done long after the vehicles have been scrapped. Most cities or states retain an accident vehicle only for a short time. In lieu of the actual vehicles, the engineer may have to rely on photographs taken by the police department or an insurance adjuster.

By applying "reverse" descriptive geometry drafting techniques, it is often possible to determine the dimensions of crush depth, impact area, and other accident parameters from photographs. For example, in a frontal view of a car that has been impacted at the side near the front, the amount of crush can be determined from the relative size of the crushed car width with respect to a headlight diameter. The known dimension of the headlight allows it to be used as a scaling factor.

Whenever possible, it is also best to visit the scene of the accident. While most investigating police officers do their best to accurately diagram the accident scene, sometimes they inadvertently omit important details. Warning sign positions and sizes, visual obstructions or shadows, brake fluid or radiator fluid pool locations, and other such details may have bearing upon the analysis of the accident events which were not noted in the accident report. Besides trying to document the accident scene, the investigating officer may also have had to direct traffic around the scene, provide first aid, write summons, locate witnesses, and handle crowds. The officer can be forgiven for missing an item that was realized to be important perhaps only months later.

H. Analysis Techniques

In assessing an accident, there are some fundamental analytical approaches that are repeatedly employed. These are the energy method, the momentum method, the dynamics method, and the kinematic method.

The *energy method* is simply the application of the first law of thermodynamics, the conservation of energy. Prior to an accident, the involved vehicles possessed kinetic energy. After the accident, the vehicles will generally have no kinetic energy left; they will be at rest. During the course of the accident events, the kinetic energy of the vehicles is dissipated in braking, skidding, plastic deformation (crushing), roll overs, changes in elevation, and other miscellaneous types of work.

In applying the energy method, the various energy dissipation events are accounted for and set equal to the initial kinetic energies of the involved vehicles. This allows the initial velocities of the vehicles to be calculated.

The *momentum method* applies the principle of the conservation of momentum. Basically, the momentum of the vehicles after impact is set equal to the momentum before impact. Allowances for inelastic impact can be made as required by determining or estimating a *coefficient of restitution*. The unknown initial velocities of the vehicles can then derived from the momentum equations.

The *dynamic method* is a straightforward application of Newton's Laws concerning mass, force and motion. Decelerating forces, both linear and rotational, are determined in order to solve for the initial velocities of the vehicles. It is typical that this method requires a greater number of detailed calculations. Often, the application of this method involves an iterative procedure where assumed forces and accelerations are revised until the assumptions "fit" with the known parameters of the accident.

The *kinematic method* uses the equations of motion exclusive of the mass and force parameters. Generally, this method uses known or assumed performance parameters for the various vehicles. This method lends itself well to graphic analysis techniques, and the resulting graphs are often easier for the non-technical person to understand.

However, while four distinct methodologies have been identified above, it is common to employ various combinations of them in an accident reconstruction analysis. One method may give a solution to a portion of the reconstruction that the others cannot provide. Thus, the application of, say, the momentum method may provide a "missing" variable that will allow the energy method to be used effectively.

Sometimes a solution obtained using primarily one methodology will be compared to the solution obtained using another. Independent convergence of two different methodologies to the same solution is sometimes taken as a verification of the solution and the underlying assumptions. However, care must be taken to insure that the two methodologies are truly independently convergent, and have not been arbitrarily "adjusted" until convergence is achieved.

CHAPTER 2

Simple Skids

A. General

When a vehicle is moving, it possesses a quantity of kinetic energy in proportion to the magnitude of its velocity and mass. In order for the vehicle to come to a stop, its kinetic energy must be dissipated. Skidding is one common way to dissipate a vehicle's kinetic energy.

In a panic situation, it is common for the driver to apply his brakes hard enough to lock the wheels. In this situation, the wheels of the vehicle do not roll; they simply lock and slide over the roadway surface. As the locked wheels slide, the tires will abrade on the roadway surface, leaving visible marks.

Skidding is a simple type of work. If no other significant dissipative effect is at work, then the kinetic energy of the vehicle is dissipated by the work done in skidding. It follows then, that if the amount of skidding is known, the amount of kinetic energy can be determined, and the initial speed of the vehicle can be calculated.

B. Basic Equations

The basic equation of engineering mechanics is Newton's Second Law:

$$F = m\,a \qquad \text{(i)}$$

where

F = the applied force,
m = the mass of the item, and
a = the acceleration.

Since the acceleration, "a", is equal to the derivative of the velocity, then equation (i) can be rewritten as follows:

$$F = (d/dt)(mv) \tag{ii}$$

where

v = velocity.

The "mv" term in equation (ii) is also referred to as the *momentum.*

By rearranging terms in equation (ii), the following equation is generated:

$$F(dt) = m(dv) \tag{iii}$$

The integration of equation (iii) is called the impulse, and is equal to the change in momentum.

$$I = \int_{t_1}^{t_2} [F(dt)] = mv_2 - mv_1 \tag{iv}$$

Equation (iv) is properly a vector equation, where the terms in the equation have both magnitude and direction. More will be done with this particular equation in later chapters.

By multiplying both sides of equation (iii) by "v", velocity, the following equation is formed:

$$Fv(dt) = mv(dv) \tag{v}$$

Integrating both sides of equation (v) produces the following:

$$\int_{t_1}^{t_2} [Fv(dt)] = \int_{v_1}^{v_2} [mv(dv)] \tag{vi}$$

$$Fv[t_2 - t_1] = ½\, m[v_2^2 - v_1^2] \tag{vii}$$

By substituting distance = velocity × time, the left term in equation (vii) becomes the standard term for work, force × distance. The right term in equation (vii) is the standard form that expresses the change in kinetic energy. Thus, the change in kinetic energy of a system is equal to the work done by or to the system.

$$F[d_2 - d_1] = \frac{1}{2} m[v_2^2 - v_1^2] \quad \text{(viii)}$$

C. The Simple Skid

A simple skid is one where the vehicle starts from some initial velocity and comes to a full stop by skidding without significant impact.

The work done by skidding is given by the following:

$$U = W f d \quad \text{(ix)}$$

where

U = the work done by skidding,
W = the weight of the vehicle
f = the coefficient of friction between the two contacting surfaces, and
d = the distance skidded.

The terms "Wf" are equal to the force necessary to push the item parallel to the ground. "W", the weight is the force normal to the ground, while "f" is a coefficient that when multiplied by "W" gives the lateral force needed to slide the item along the ground.

Since the initial velocity of the vehicle is some unknown "v", and the final velocity of the vehicle in this case would be zero, then equation (viii) can be rearranged as follows:

$$\frac{1}{2} mv^2 = W f d \quad \text{(x)}$$

Solving for "v", the initial velocity before skidding began, then equation (x) becomes:

$$v = [2(W/m)fd]^{1/2} \quad \text{(xi)}$$

Since the weight of an object divided by its mass is equal to the gravitational acceleration, then equation (xi) is further simplified to the following:

$$v = [2 g f d]^{1/2} \quad \text{(xii)}$$

where

g = 32.17 ft/sec^2 or 9.81 m/s^2

Equation (xii) is a version of the common "skid formula" which is taught to police officers. For police use, a nomograph is often provided for easy solution of "v".

Figure 1. Two sets of skids made by truck dual wheels. Note traffic light where skids end.

It is interesting to note that the weight or mass of the vehicle cancels out of the equation in this simple case. Calculation of the initial velocity of the vehicle depends solely on the length of the skid and the coefficient of friction.

The following table lists some typical friction coefficients for calculating forward motion skidding work.

Table 1. Some Common Frictional Coefficients For Car Tires on Roadways

Surface Type	f-value
gravel and dirt road	0.35
wet, grassy field	0.20
dry asphalt	0.65
wet asphalt	0.50
dry concrete	0.75
wet concrete	0.60
snow covered	0.20 – 0.25
ice	0.10 – 0.15
loose moist dirt that allows tire to sink about 2" (5 cm)	0.60 – 0.65

Figure 2. "Classic" skid marks: tapered beginning and abrupt end where vehicle stopped.

D. Tire Friction

The figures provided in Table 1 are reasonably accurate for relatively short skids. However, it is known that the coefficient of friction is not wholly constant during a skid. As the skid progresses, the coefficient of friction will decrease as a function of the forward velocity. This decrease in the coefficient of friction can be approximated by the following linear equation:

$$f = f_1 - cv \qquad \text{(xiii)}$$

where

f = coefficient of friction,
f_1 = initial coefficient of friction term (static friction),
c = constant determined by experimentation, and
v = forward velocity.

The following table lists some general values for solving equation (xiii) for automobile tires.

Table 2. Typical Values for Solution of "f"

conditions	f_1	c
dry concrete	0.85	0.0017
wet concrete	0.70	0.0034

Note: the above values are to be used when "v", velocity, is given in ft/sec units.

Figure 3. Skipping skid marks are usually the result of light load in the trailer or when trailer brakes alone are applied. Measure the whole length; do not omit the spaces in the "skips."

A comparison of the constant values for "f" given in table 1 with the linear equation function for "f" finds that they are equal to one another when the velocity of the car is 40 mph on dry pavement. For wet concrete pavement, the values are the same when the velocity of the car is 20 mph.

In addition to varying with forward velocity, in trucks the road-to-tire coefficient of friction will vary with truck loading. As the loading on the tire is increased, the coefficient of friction decreases. The following equation approximates that effect:

$$f = f_1 - 0.22(L/N) \tag{xiv}$$

where

f_1 = initial coefficient of friction term (static friction),
L = load carried by the tire, and
N = rated load of the tire at the proper inflation pressure.

While the factor "f_1" is essentially the same as used before from Table 2, it is worth noting that truck tires typically have a coefficient of friction about 10% greater than passenger car tires. This is due to a different rubber blend used in truck tires.

The two factors, load and velocity, can be combined into a single equation for trucks as follows:

$$f = f_1 - cv - 0.22(L/D). \tag{xv}$$

Figure 4. Differences in darkness of skid marks are due to load, tire wear, braking and road geometry.

Because road-to-tire friction can be affected by excess dirt, chemical films, surface polishing by traffic wear, moisture, the brand and size of tires, and other miscellaneous factors, it is important to recognize that the "f" values may exhibit significant variation even along the same area of roadway. In a four-lane highway, the frictional coefficient of the inside lane may even vary somewhat from the outside lane.

When very accurate values of "f" are required, it may be necessary to obtain the values experimentally. This can be done by:

- renting a car similar to the one in question, deliberately skidding it at various initial speeds, measuring the skid lengths, and calculating "f." For obvious reasons, this method is not generally recommended for amateur drivers.

- pulling a weighted sled with a fixed wheel and tire, and measuring the draw of the sled. The ratio of the draw to the sled weight gives a value of "f."

- towing a similar car with the rear wheels locked and measuring the draw. Because the front end is lifted, the "f" value is the ratio of draw to the weight on the rear tires (or the weight of the car minus the lifted weight at the tow hoist).

E. Multiple Surfaces

In some instances, a vehicle may skid over several different surfaces in succession. This could happen, for example, if the vehicle began skidding on the roadway, skidded on the shoulder, and then skidded on the dirt alongside the roadway.

In such cases, the skid trajectory is simply divided into segments according to the surface type. Each segment is a work term, and the sum of the work terms is set equal to the initial kinetic energy of the vehicle.

With multiple surfaces, equation (x) is modified as follows:

$$(1/2)mv^2 = (W)\sum_{i}^{n}(f_i d_i) \qquad \text{(xvi)}$$

Solving for "v" in equation (xvi) gives the following:

$$v = [2g(f_1 d_1 + f_2 d_2 + \ldots\ldots.. + f_n d_n)]^{1/2} \qquad \text{(xvii)}$$

In multiple-surface skids, it is sometimes useful to set each skid segment work term equal to the change in kinetic energy of the vehicle during that skid segment. In this fashion, the various individual velocities during the overall skid trajectory can be calculated.

For example, assume that a vehicle skidded first on pavement and then on soil. The skid trajectory would have two segments, and correspondingly, two skid work terms. Let the initial velocity of the vehicle and the start point of the pavement skid be designated by subscript 1. Let the end point of the pavement skid and the beginning of the soil skid be designated by subscript 2. Then let the end point of the soil skid, where the car finally stopped, be designated by the subscript 3. Thus, the following equations would describe the situation:

$$v_1 = [2g(f_p d_p + f_s d_s)]^{1/2}$$

$$v_2 = [v_1^2 - 2gf_p d_p]^{1/2} = [2gf_s d_s]^{1/2}$$

$$v_3 = 0$$

With all three velocity terms solved, a plot of "v" versus distance can be constructed. From such a plot, intermediate velocities at any point along the skid trajectory can be picked off. Such plots are very useful in vehicle-pedestrian accidents. If the location of the impact with the pedestrian is known from pavement marks or other indicators, its location can be measured from one end of the skid. From this measurement, the velocity of the vehicle at the point of impact with the pedestrian can be determined.

(Note: In accidents involving pedestrians and cars or heavier vehicles, it is usually assumed that the impact with the pedestrian has no significant effect on the trajectory of the vehicle. This may not be true in a motorcycle impact, where the weight ratios are closer.)

F. Calculation of Skid Deceleration

Occasionally, it is useful to know the deceleration (i.e., negative acceleration) for a given skid. In this respect, the deceleration of the vehicle by skidding is simply the frictional force opposing travel divided by the mass of the vehicle.

The opposing friction force is given by:

$$F_f = W\,f \qquad \text{(xviii)}$$

where

F_f = friction force opposing travel,
W = weight of the vehicle, and
f = coefficient of friction.

Dividing both sides of equation (xviii) by the mass of the vehicle, the resulting term will be the deceleration of the skid.

$$a_f = g\,f \qquad \text{(xix)}$$

In police nomenclature, the above term may be called the "drag" of the vehicle. The term "drag" refers to any frictional force that slows the vehicle.

G. Speed Reduction by Skidding

Sometimes skidding will not last long enough to bring the vehicle to a stop. In such cases, the skidding simply reduces the initial velocity of the vehicle to some lower value. If either the initial velocity just prior to skidding, or the velocity just after skidding is known, the other can be calculated.

$$\tfrac{1}{2}\,m[v_2^2 - v_1^2] = W\,f\,d \qquad \text{(xx)}$$

$$v_2 = [2\,g\,f\,d + v_1^2]^{1/2}$$

$$v_1 = [v_2^2 - 2\,g\,f\,d]^{1/2}$$

H. Some Considerations of Data Error

In reviewing equation (xii), it is noted that only two terms are subject to possible error: "f" and "d". Since the length of the skid mark is usually measured with reasonable precision with a tape measure, the greatest apparent source of error is then the estimation of "f".

However, due to the particular construction of the skid equation, that is equation (xii), the error is actually smaller than one might suppose. This is due to the square-root effect of decreasing the error range.

To illustrate this point, suppose a vehicle skidded 100 ft. and the estimated coefficient of friction is 0.70. The calculated initial velocity is then 67.1 ft/sec. Assuming that the skid measurement is accurate to the nearest foot, if the "f" value is increased 14% to 0.80, then the initial velocity calculates to be 71.7 ft/sec, an increase of 7% from the previous figure. Similarly, if the "f" value is decreased 14% to 0.60, the calculated initial velocity is then 62.1 ft/sec, a decrease of 7%.

Thus, the amount of error caused by either an imprecise estimation of "f" or measurement of "d" does not cause the same amount of error in the calculation of the initial velocity. The error in the calculation of initial velocity is a function of the square root of the combined error of the "f" and "d" terms.

For example, if the "df" term is in error by +21%, the resulting initial velocity calculation will only be in error by 10%. Similarly, if the "df" term is in error by -19%, the resulting initial velocity calculation will only be in error by 10%. Thus, contrary to intuition, the final answer can be more accurate than the data upon which it relies.

I. Curved Skids

Work and energy are both scalar quantities. They have magnitude but not direction. In vector algebra, work is defined as the dot product of force and direction.

$$\mathbf{F} \cdot \mathbf{d} = Fd(\cos \alpha) = U \tag{xxi}$$

where

- U = work (a scalar),
- $\mathbf{F}$ = force (a vector),
- $\mathbf{d}$ = distance (a vector)
- α = the acute angle between the force and the distance.

If a skid curves or otherwise does not follow a straight line, the skid should be measured along the curved line. If a vehicle spins while it skids, the length of the skid should be measured along the curves and loops of the skid to obtain the total work of the skid.

This is because energy is dissipated by the tire sliding on the pavement in the direction of the slide. It takes work to push forward a vehicle with locked wheels. It also takes work to spin a vehicle with locked wheels even if the vehicle does not end up moving forward any distance.

Thus, to be precise in the amount of energy dissipated by skidding, the curvilinear length of the skid should be used rather that the start-to-finish straight-line distance. If is often the case that the lengths of skid marks noted in police reports are given as the straight line distance from where the skid began to where it stopped. When necessary, allowances should be make to account for any nonlinear skid mark distances.

J. Brake Failures

Occasionally, the brakes on a vehicle will fail prior to usage, or just when skidding occurs (and the internal hydraulic pressure is at a maximum). In such cases, only two of the four wheels will lock up and skid. This is because most automobile brake systems have dual hydraulic systems or loops. Thus, when a leak in the hydraulic brake system occurs, only half of the system becomes inoperative.

In such cases where there has been a failure in the hydraulic brake system, the vehicle will be decelerated by only two of the four wheels. Thus, the weight distribution of the vehicle becomes important.

If the weight in a four-wheeled vehicle were evenly distributed, each wheel would support one-fourth of the weight of the vehicle. The sum of the work done by each wheel in decelerating the vehicle would then be as follows:

$$U = (1/4)W \sum_{i=1}^{n=4} (f_i d_i) \qquad \text{(xxii)}$$

where the subscripts 1...4 indicate each of the wheels on the vehicle.

If all the distances skidded by the individual wheels are the same, and all the coefficients of friction are the same, equation (xxii) reduces to the standard skid formula.

However, most cars and trucks do not have an equal distribution of weight among the wheels. In cars with a front-mounted engine, it is common for the front two wheels to carry 65% of the total curb weight of the vehicle. (Curb weight is the amount the car weighs with no passengers, freight or cargo.) In such a car then equation (xxii) would become:

$$U = W\,f\,[0.325d_1 + 0.325d_2 + 0.175d_3 + 0.175d_4] \qquad \text{(xxiii)}$$

where the subscripts 1 and 2 are for the front wheels, and 3 and 4 are for the rear wheels. The above also assumes that the term "f" is the same for all wheels.

If one or more of the wheels were on a different surface, then it might also be necessary to assign a specific "f" value for each wheel.

If the hydraulic brake system is one where the front right wheel is on the same hydraulic circuit as the left rear wheel, and vice versa, then the work done by skidding will be done by the two wheels, front and back, of the remaining operable hydraulic circuit. Applying equation (xxiii), then the work done by the partially disabled brake system would be as follows:

$$U = W\,f\,[0.325\,d_1 + 0.175\,d_3] \qquad \text{(xxiv)}$$

If "d_1" = "d_3", then equation (xxiv) reduces to the following:

$$U = \tfrac{1}{2}\,W\,f\,d \qquad \text{(xxv)}$$

Similarly, if the brake system were designed so that both front brakes and both rear brakes were on independent hydraulic circuits, the work done in a skid by either pair of brakes would be as follows:

$$U = (0.65)\,W\,d\,f \quad \text{for front wheels} \qquad \text{(xxvi)}$$

and

$$U = (0.35)\,W\,d\,f \quad \text{for rear wheels} \qquad \text{(xxvii)}$$

Equation (xxvii) also applies if the emergency brake is used. It is customary for the emergency brake to engage the two rear wheel brakes. The emergency brake will normally be actuated by a direct mechanical cable linkage to the two rear-wheel brakes.

For unloaded pickup trucks, and for other kinds of commercial vehicles, the weight distribution may favor the front end even more. Certain vehicles, like the Volkswagen "bug" or fully loaded commercial vehicles, may be heavier in the rear than in the front. The weight distribution of each vehicle must be individually assessed.

In general, most vehicles will have their GAWR (gross axle weight ratings) for both the front and rear axles posted on the certification label, which is required by law. This label is usually found on the driver's door post of passenger cars. The GAWR-F (front axle) and the GAWR-R (rear axle) should equal the total vehicle GVWR (gross vehicle weight rating). The gross axle weight rating lists the maximum total load which that axle can carry safely. This includes the weight of the vehicle on the axle and any other load being carried.

When the appropriate work term is determined, the value for "U" is then set equal to the kinetic energy term, and the initial velocity of the vehicle is solved as before.

The preceding analysis has assumed that the load distribution between the front and rear wheels follows the static load distribution. For most purposes, this assumption is reasonably accurate. However, during hard braking there is a dynamic effect which modifies this assumption. This is discussed in section L.

K. Changes in Elevation

In all of the previous analyses, it has been assumed that there was no change in elevation from where the skidding started to where it ended. In cases where there is a significant change in elevation, a potential energy term must be added to the energy balance. Intuitively, it is reasonable to expect that a vehicle will skid farther in a downhill direction than uphill.

In previous equations, it was generally assumed that the kinetic energy of the vehicle was wholly converted into work done by skidding, that is:

$$\Delta KE = U \tag{xxviii}$$

where

KE = the kinetic energy of the vehicle, and
U = the work done by skidding.

To account for changes in elevation during skidding, a potential energy term is added to equation (xxviii).

$$\Delta KE + \Delta PE = U \tag{xxix}$$

Potential energy is often described as stored energy, such as exists, for example, when a spring is compressed. Work is done to compress the spring, and the work can be recovered as the spring is released. During the time that the spring is compressed, the work is stored in the spring as potential energy.

When a stone is dropped, it is accelerated by gravity. Ignoring air friction, the velocity of the stone after being dropped is given by:

$$v = g\,t \tag{xxx}$$

where

v = the velocity after time "t" has elapsed, and
g = the gravitational acceleration.

By substituting "d(x)/dt" for "v" and integrating, the following equation relating the distance the stone has fallen to the elapsed time is obtained.

$$h = \frac{1}{2} g t^2 \quad \text{(xxxi)}$$

where

h = the height fallen by the stone.

Substituting "$v = gt$" into equation (xxxi) and multiplying both sides of the equation by "g", equation (xxxi) is transformed into the following:

$$g h = \frac{1}{2} v^2 \quad \text{(xxxii)}$$

By further multiplying both sides of equation (xxxii) by the mass "m", the following equation is obtained:

$$m g h = \frac{1}{2} m v^2 \quad \text{(xxxiii)}$$

The left term of equation (xxxiii) is the expression for the change in potential energy due to a change in elevation. The right term is the standard term for kinetic energy. Thus, when a stone is dropped, its potential energy is converted to kinetic energy. Conversely, when a stone is elevated, its kinetic energy is decreased, and its potential energy is increased.

If a vehicle skids downhill, the change of potential energy is converted into additional kinetic energy that must be dissipated by the skid: the skid will be longer. Similarly, if a vehicle skids uphill, kinetic energy is converted into potential energy: the skid will be shorter. The following equation describes this effect.

$$\frac{1}{2} m [v_2^2 - v_1^2] + m g[h_2 - h_1] = W f [d_2 - d_1] \quad \text{(xxxiv)}$$

If it is assumed that the vehicle comes to a stop, i.e. "$v_2 = 0$", equation (xxxv) simplifies to the following:

$$v = [2 g (d f - (h_2 - h_1))]^{1/2} \quad \text{(xxxv)}$$

A quick way to determine the relative effect of an elevation change upon the calculation of initial velocity is to apply the following:

$$\Delta = 1.00 - \{[f\,d - h_2 + h_1] / f\,d\}^{1/2} \qquad \text{(xxxvi)}$$

Obviously, if the change in elevation is negligible with respect to the "fd" term, then the elevation terms can be dropped and simpler equations can be applied with no significant loss of accuracy.

Most police reports do not document changes in roadway elevation unless the vehicle has gone over a cliff or careened into a deep ditch. Generally, an assessment of elevation change will have to be done by on-site inspection of the accident area.

Occasionally, roadway topographical drawings may be available from the local unit of government responsible for the maintenance of the roadway in question. However, sometimes these drawings are out of date, and may not document recent repairs or modifications that may affect an accurate assessment of elevation in the accident area.

L. Load Shift

When a vehicle skids, the actual load carried by the tires can change. This is due to the different forces present during braking as opposed to when the vehicle is at rest. The change in load distribution is observable as "brake squat", where the front end of the car tilts down and the rear end tilts up as the car brakes. (A similar but reversed "squat" effect occurs during acceleration where the front end raises up and the rear end drops down.)

When a car is at rest, the usual static equation applies to the loads carried by the front and rear tires.

$$W = Q + R \qquad \text{(xxxvii)}$$

where

W = weight of the car,
Q = load carried by the rear tires, and
R = load carried by the front tires.

Let "b" be the distance from the front axle to the rear axle, and "a" be the distance from the center of gravity of the car to the front wheel axle. The moment equation about the front axle is then as follows:

$$Q(b) - W(a) = 0 \qquad \text{(xxxviii)}$$

In the above moment equation, the specific point of rotation is considered to be where the front wheels contact the road. By combining the above two equations, the front and rear load distributions can be determined.

$$Q = W\,[a/b] \qquad \text{(xxxix)}$$

$$R = W\,[1 - (a/b)]$$

Of course, these same two equations can be used in a reverse sense to calculate the center of gravity of a vehicle when the rear and front loads and the weight of the vehicle are known.

In the above analysis, it was assumed that all the forces are pointing up or down. However, when the vehicle is being braked, there is a forward directed force "F". This force is the mass of the vehicle times the braking acceleration (or deceleration), and is directed through the center of gravity of the vehicle parallel to the ground (i.e., horizontally). This force is given by the following expression:

$$F = m\,a = m\,g\,f = W\,f \qquad \text{(xl)}$$

where

f = coefficient of friction.

The inclusion of "F" in the moment equation (xxxviii) modifies it to the following:

$$Q(b) - W(a) + F(h) = 0 \qquad \text{(xli)}$$

$$Q(b) - W(a) + Wfh = 0$$

where

h = the height of the car's center of gravity from the ground.

The loads carried by the front and rear wheels then change to the following:

$$Q = W[(a\ /\ b) - f(h\ /\ b)] \qquad \text{(xlii)}$$

$$R = W[1 - (a\ /\ b) + (f\,h\ /\ b)]$$

As the equations show, the load on the rear wheels lightens and the load on the front end increases while the car is skidding and decelerating. Assuming that the car has a load distribution of 60% in front and 40% in the rear while at rest, then "a/b" in the above equation is 0.40. Further, assuming that "f = 0.70" and "h/b = 0.20", then skidding will cause the load in the rear to lighten by 14% and the load in the front to increase by 14%.

In general, the load shift caused by skidding does not affect skid calculations. However, it may become important when partial brake system failures occur, or when the emergency brakes are used in lieu of normal braking. When either condition occurs, the term "Wf(h/b)" has to be modified to agree with the reduced braking force.

For example, when only the rear brakes are applied, the following equations indicate the load shift during skidding:

$$Q = W(a / b)[1 - (f h / b)] \quad \text{(xliii)}$$

$$R = W[1 - (a / b) + (a / b)f(h / b)]$$

Using the same assumptions as before (f = 0.70, a/b = 0.40, h/b = 0.20), we find that skidding will cause a load shift of only 5.6% when only the emergency brakes are used. This is because of the reduced braking capacity of the emergency brakes as compared to the full system.

M. Anti-Lock Brake Systems (ABS)

It might be assumed that anti-lock brakes would eliminate skidding and skid marks. This is not quite the case. While it is true that anti-locking brakes do not leave a heavy black skid mark like conventional brake systems that can lock up the wheels, a light skipping or alternating pattern on the pavement can still be discerned in most cases.

Anti-lock brake systems are generally designed to release the wheels to allow them to roll and not fully lock up. While this reduces the sliding or slip between the tire and the road, it does not fully eliminate it. In fact, many anti-lock braking systems are calibrated to release when slip just exceeds 30%. Slip is defined as follows:

$$\text{Slip} = [1 - v_t/v] \quad \text{(xliv)}$$

where

v = forward velocity of the vehicle, and

v_t = tangential velocity of the wheel where it contacts the road.

When slip is 100%, the vehicle is skidding; when slip is 0%, the vehicle has full traction between the tire and the roadway. Between

15% and 30% slip on dry pavement, the maximum braking effect is achieved with minimal sideslip. On wet pavement, maximum braking effect with minimum sideslip occurs when slip is between 30% and 60%.

Thus, even when anti-lock brake systems are employed, there is still some slippage between the tire and the road surface. It often occurs in a skipping pattern because the anti-lock system will alternately open and release the brakes as the slip set point is exceeded. Because the skid pattern of anti-lock brakes is different in appearance from conventional braking skid marks, it is often overlooked at the accident scene. In some cases, once it was learned that a car involved in the accident had ABS brakes, it was automatically assumed that there were no skid marks; and therefore no search for marks was conducted!

In general, a full skid with locked wheels will bring a car to a stop in the shortest distance. However, during a full skid, the driver has little control over the vehicle. With ABS brakes, the actual braking distance may be slightly longer, but steering control can still be maintained.

On dry pavement, the same calculation methods for locked brake skids are applicable to ABS type skidding. On wet pavement, the same calculation methods can be used also, but the coefficient of friction should be considered to be about 10% greater.

Example Problems

Problem 1

A car was traveling eastbound in an area where the speed limit was posted at 35 mph. The car approached a four-way intersection and apparently attempted to get across the intersection before the traffic light turned red. However, before the car could get across the intersection, a pickup truck approaching from the south end of the intersection made a right turn on a red light, and turned into the same lane as the car. The driver of the car slammed on her brakes and skidded to a stop missing the pickup truck.

The officer at the scene noted that the skid marks were 54 feet long, and that the pavement was dry concrete. Was the car traveling at a legal speed?

Solution

In Table 1, the coefficient of friction for dry concrete is given as 0.75. Applying equation (xii), the velocity of the car when the skidding initiated is found as follows:

$$v = [2\ f\ g\ d]^{1/2}$$

$$v = [2(0.75)(32.17\ ft/sec^2)(54\ ft)]^{1/2}$$

$$v = [2{,}606\ ft^2/sec^2]^{1/2}$$

$$v = 51\ ft/sec = 35\ mph.$$

Thus, when the driver of the car slammed on her brakes, she was traveling at the speed limit.

Problem 2

A car traveling at an unknown speed skidded and struck a pedestrian in a crosswalk. The weather was poor and there was packed snow on the streets, which obscured the crosswalk markings.

From marks found on the roadway surface, it was found that the car slid or skidded 64 feet prior to striking the pedestrian, and an additional 15 feet afterwards. A police car tested the street by driving at 20 mph and slamming on the brakes. It was noted that the police car slid 70 feet. How fast was the car traveling when it hit the pedestrian?

Solution

The total skid length was 79 feet. Since striking a pedestrian has little effect on the speed of a full-sized car, the effect of the pedestrian can be neglected without loss of accuracy.

By rearranging the terms of equation (xii), and using the information obtained when the police car was tested, the coefficient of friction "f" can be found.

$$f = (v^2)/\ 2\ g\ d$$

$$f = (29.3\ ft/sec)^2/(2)(32.17\ ft/sec^2)(70\ ft)$$

$$f = 0.19$$

Now applying equation (xii) but using only the last 15 feet of the car's trajectory, the distance covered after impact with the pedestrian, the speed at the time of impact with the pedestrian is obtained.

$$v = [2\ g\ f\ d]^{1/2}$$

$$v = [(2)(32.17\ ft/sec^2)(0.19)(15\ ft)]^{1/2}$$

$$v = 13.6\ ft/sec = 9.2\ mph$$

The speed of the car when sliding or skidding began is given as follows:

$$v = [(2)(32.17\ ft/sec^2)(0.19)(79\ ft)]^{1/2}$$

$$v = 31.1\ ft/sec = 21.2\ mph$$

Problem 3

A car skidded first 30 feet along the asphalt roadway, then skidded 12 feet along the shoulder, and then fell from a precipice. The roadway was dry asphalt. The shoulder was firm dirt. If analysis of the fall indicates that the vehicle was traveling forward at 35 mph when it went over the precipice, how fast was it going when skidding first began?

Solution

In this case there are two skid surfaces: dry asphalt and firm dirt. For the asphalt, a value of 0.60 will be used for "f". For the firm dirt, an "f" value of 0.55 will be used.

Let v_3 = 35 mph = 51.3 ft/sec

Applying equation (xx), the following is obtained:

$$v_2 = [2\ g\ f_2 d_2 + v_3^{\,2}]^{1/2}$$

$$v_2 = [2(32.17\ ft/sec^2)(0.55)(12\ ft) + (51.3\ ft/sec)^2]^{1/2}$$

$$v_2 = 55.3\ ft/sec = 37.7\ mph$$

$$v_1 = [2gf_1d_1 + v_2^{\,2}]^{1/2}$$

$$v_1 = [2(32.17 \text{ ft/sec}^2)(0.60)(30 \text{ ft}) + (55.3 \text{ ft/sec})^2]^{1/2}$$

$$v_1 = 64.9 \text{ ft/sec} = 44.3 \text{ mph.}$$

Thus, when skidding started, the car was traveling at 44.3 mph. When the car left the roadway and onto the shoulder, it was traveling at 37.7 mph. Skidding along the shoulder then reduced its speed to 35 mph, when it went over the edge.

Problem 4

A car experienced a total failure of its hydraulic brake system. In order to effect deceleration and stop, the driver applied the emergency brake by pulling the brake lever. Compare the distance the car would skid using only the emergency brake with the distance it would skid with the unfailed hydraulic brake system. Compare also the relative decelerations.

Solution

By rearranging equation (xii) to solve for distance, the following is obtained:

$$d = (v^2)/(2 \text{ g f})$$

Assuming that the front of the car carries 65% of the load of the car, then for the emergency brake skid,

$$d_e = (v^2)/(2g)(0.35)(f)$$

If a ratio is made between the total skid distance using only the emergency brakes versus the total skid distance using the car's regular brakes, the following is obtained:

$$d_e/d = 1.0/0.35 = 2.86$$

Thus, the car will skid 2.86 times further using only the emergency brakes than when using its regular brake system.

Applying equation (xix) to obtain the decelerations, it is found that the deceleration using the regular braking system is 0.7g or 22.5 ft/sec^2, and the deceleration using only the emergency brakes is (0.7)(0.35)g or 7.88 ft/sec^2.

Problem 5

A car skids a total of 58 feet on wet concrete pavement. Measurements at the scene show that the roadway has a slope of 5% in the area of the skid and that the coefficient of friction is 0.60. What is the effect of the change in elevation of the roadway slope on the calculation of the initial velocity? Consider the slope first as being with the skid, and then against the skid.

Solution

Applying equation (xii) to find the initial speed on level ground, it is found that:

$$v = [2(32.17\ \text{ft/sec}^2)(58\ \text{ft})(0.60)]^{1/2} = 47.3\ \text{ft/sec} = 32.3\ \text{mph}.$$

A 5% grade means that for every 100 feet of horizontal distance, the roadway will elevate 5 feet. Because the skid was 58 feet long, then the change in elevation along the skid length would have been:

$$h = 0.05(58\ \text{ft}) = 2.9\ \text{ft}$$

•Note: due to the small angle involved, the skid distance along the roadway is taken to be the same as along a horizontal.

Applying equation (xxxvi) it is found that the effect of elevation change on the initial velocity would be as follows:

$$\Delta = 1.00 - \{[0.60(58\ \text{ft}) \pm (2.9\ \text{ft})]/[(0.60)(58\ \text{ft})]^{1/2}$$

$$\Delta = \pm\ 4\%$$

Thus, the change in elevation in this instance would make a difference of 4% in the calculation of the initial velocity. The calculated initial velocities would be then:

v = 45.3 ft/sec = 30.9 mph with the vehicle skidding downhill, and

v = 49.3 ft/sec = 33.6 mph with the vehicle skidding uphill.

Problem 6

A car skids 74 feet and comes to a stop. The coefficient of friction is estimated as being between 0.6 and 0.7. The area has insignificant change in elevation. What is the range of error in the calculation of initial speed of the car assuming that the accuracy of skid measurements is ± 1 ft?

Solution

The maximum initial speed is given by the following:

$$v_{max} = [2(32.17 \text{ ft/sec}^2)(0.70)(75 \text{ ft})]^{1/2} = 58.12 \text{ ft/sec.}$$

The minimum initial speed is given by the following:

$$v_{min} = [2(32.17 \text{ ft/sec}^2)(0.60)(73 \text{ ft})]^{1/2} = 53.09 \text{ ft/sec.}$$

Thus, the initial velocity of the car is calculated to be 55.6 ft/sec with a range of ± 2.52 ft/sec, or an error range of ± 4.5%. Incidentally, the error range in the skid work term inside the brackets is ± 9.9%.

CHAPTER 3

Simple Falls

A. General

In some instances, vehicles will leave the roadway and go over a cliff, drop into a ditch, or otherwise free-fall from a precipice. In the reconstruction of such an accident, it is important to determine the velocity of the vehicle just prior to the fall. In general, if the height of the fall is known and the horizontal travel is known, the velocity just prior to fall can be calculated.

B. Basic Equations

The motion of an object during fall consists of two independent motions: one in the vertical or "y" direction, and the other in the horizontal or "x" direction.

As was discussed in section K of chapter 2, when an object falls, some of its potential energy due to elevation in a gravitational field is converted to kinetic energy.

$$PE_2 - PE_1 + KE_2 - KE_1 = 0 \tag{i}$$

Equation (i) indicates that the energy change associated with a change in elevation of an object is wholly converted into a change in kinetic energy of the object. More significantly, equation (i) is restated as follows:

$$W[h_2 - h_1] = \frac{1}{2} m[v_2^2 - v_1^2] \tag{ii}$$

where

W = weight of the object,

m = mass of the object,
h_2 = height of the point of impact after the fall,
h_1 = height of the point where fall began,
v_2 = velocity at point of impact after the fall, and
v_1 = velocity when fall began.

Since most vehicles do not initially have any significant vertical velocity at the point just prior to fall, "v_1" is zero. Also, since "h_2" will be the low point, it is usually considered to be zero also. Equation (ii) can thus be modified and reduced to the following:

$$v = [2\,g\,h]^{1/2} \tag{iii}$$

where
v = the vertical velocity at impact, and
h = the vertical distance the vehicle fell.

Because the acceleration of gravity is constant for such situations, the vertical velocity at impact can also be calculated by the following:

$$v = g\,t \tag{iv}$$

where t = the time of fall.

By substituting "$v = dy/dt$" and rearranging terms, the following is obtained from equation (iv):

$$dy = gt(dt) \tag{v}$$

Integrating equation (v) with respect to time and letting "$y_2 - y_1 = h$", the following relation is derived:

$$h = \tfrac{1}{2}\,g\,t^2 \tag{vi}$$

where
t = time elapsed during the fall.

Of course, while the vehicle was falling in the vertical direction, it was still moving horizontally. In fact, neglecting air resistance, the vehicle would continue to move horizontally at the same velocity it had before the fall, and would continue to do so until impact with the ground (or whatever) occurs. The amount of time that the vehicle will continue to travel horizontally will be the same "t" found in equation (vi) above. Solving for "t" in equation (vi) gives:

$$t = [2h/g]^{1/2} \quad \text{(vii)}$$

Let "x" be the horizontal distance traveled by the vehicle from the point of "take off" to the point of impact. Then the velocity of the vehicle just prior to the fall is given by:

$$u = x/t = x[2h/g]^{-1/2} = x[g/2h]^{1/2} \quad \text{(viii)}$$

where

u = the horizontal velocity prior to fall.

The velocity at impact, which includes both the horizontal and vertical velocity components, is given by the following vector equation:

$$V = (u)i + (v)j \quad \text{(ix)}$$

where i and j are the horizontal and vertical unit vectors.

In absolute terms, the total velocity is given by:

$$V = [u^2 + v^2]^{1/2} \quad \text{(x)}$$

The angle of impact is then:

$$\alpha = \arctan(v/u) \quad \text{(xi)}$$

where α is measured from the impact point and upwards.

A vehicle with no forward horizontal motion would have an angle of impact of 90 degrees.

C. Ramp Effects

If the take off point is not level, an additional upward or downward velocity component must be taken into consideration. Let "ϕ" be the acute angle that the vehicle's take off ramp makes with the horizontal. Then, the velocity of the vehicle can be separated into orthogonal components as follows:

$$v = V \sin \phi \quad \text{(xii)}$$

$$u = V \cos \phi$$

where

V = the velocity of the vehicle at take off in the direction of travel.

Assuming that "v" is upward and is considered positive, then the vertical motion equation is as follows:

$$v = V \sin \phi - g t \tag{xiii}$$

The highest elevation the vehicle will attain during its trajectory is calculated as follows:

$$t_a = V(\sin \phi)/g \tag{xiv}$$

$$y_a = \tfrac{1}{2} g t_a^2 = \tfrac{1}{2} V^2(\sin^2 \phi)/g$$

where

t_a = the time it takes to get to the top of the arc in the trajectory and

y_a = vertical distance from the take off point to the top of the arc.

The total fall of the vehicle will then be the height of the ramp from where the vehicle impacted plus the height it attained during its trajectory above the end of the ramp. The total time of fall will include the time taken to reach the apex of the arc plus the fall time from that same point. Let the total height of fall, from the apex of the arc to the point of impact, then be as follows:

$$H = h + y_a \tag{xv}$$

$$H = h + V^2(\sin^2 \phi)/(2g) = h + \tfrac{1}{2} g t_a^2$$

The time needed to fall from the apex of the trajectory arc to the point of impact is given by the following:

$$t_b = [2H/g]^{1/2} \tag{xvi}$$

The total time of the fall then, from the take off point, through the upward trajectory arc, and down to the point of impact is then as follows:

$$t = t_a + t_b = V(\sin \phi)/g + [2H/g]^{1/2} \tag{xvii}$$

$$t = V(\sin \phi)/g + [2h/g + V^2(\sin^2 \phi)/g^2]^{1/2}$$

As before, the horizontal speed of the vehicle, "u", can be calculated by the following:

$$u = V(\cos\phi) = x/t \qquad \text{(xviii)}$$

If equation (xviii) is substituted into equation (xvii), the following is derived:

$$x = V^2(\cos\phi)(\sin\phi)/g + V(\cos\phi)\{2h/g + V^2(\sin^2\phi)/g^2\}^{1/2} \qquad \text{(xix)}$$

Equation (xix) gives the horizontal distance "x" travelled by the vehicle in terms of the initial velocity, the angle of the ramp, and the height from the ramp to the point of impact. As would be expected, when the angle "ϕ" is zero, equation (xix) reduces to the following:

$$x = V[2h/g]^{1/2} \qquad \text{(xx)}$$

or

$$u = x[2h/g]^{-1/2}$$

Equation (xx) is, of course, a restatement of equation (viii), which is used when the precipice is flat, that is, there is no ramp angle.

When there is no drop off, and the vehicle is simply catapulted by driving up and over a ramp, equation (xix) reduces to the following:

$$x = 2V^2(\cos\phi)(\sin\phi)/g \qquad \text{(xxi)}$$

which is the distance travelled from the end of the ramp to the point of impact, which occurs at the same elevation as the end of the ramp. The elevation above the end of the ramp attained during such a trajectory is given by equation (xiv).

In general, equation (xix) is solved for "V" by an iterative process. An initial guess for "V" is selected, and the equation is solved for "x". This "x" is then compared to the measured value for "x", the horizontal distance traveled during fall by the vehicle. A new value of "V" is then selected and the process is repeated until the calculated "x" matches the measured "x".

In cases where the ramp is inclined at a negative angle, that is, an angle that dips below the horizontal, a slightly different derivation is required. With the ramp pointing downward, then:

$$u = V\cos(-\phi) = V\cos\phi \qquad \text{(xxii)}$$

$$v = V(\sin(-\phi) = -V \sin \phi$$

It follows then that:

$$h = V(\sin \phi)t + \tfrac{1}{2} g t^2 \qquad \text{(xxiii)}$$

The solution of the time of fall, "t", in equation (xxiii) involves a quadratic equation in which there are two possible values of "t".

$$t = -V(\sin \phi)/g \pm [V^2(\sin^2 \phi)/g^2 + 2h/g]^{1/2} \qquad \text{(xxiv)}$$

Except for unusual cases, the positive sign is used in front of the second term of equation (xxiv).

As before, the horizontal velocity is determined by dividing the horizontal distance by the lapsed time, "t".

$$u = V \cos \phi = x/t \qquad \text{(xxv)}$$

where in this case, "t" is obtained from the solution of equation (xxiv).

The method of solution for this situation is as follows:

- Set the "t" in equation (xxiv) equal to "$x/V(\cos \phi)$".
- Choose a good guess for "V" and solve for "x".
- Appropriately alter "V" iteratively until the calculated "x" matches the actual "x".

When the above procedure is followed, it yields a solution for "x" similar to equation (xix), except that the first term is negative, while in equation (xix) it is positive. In fact, equation (xix) is appropriate for both situations, an uphill ramp and a downhill ramp, if the appropriate positive and negative sign conventions are observed with respect to the ramp angle.

D. Air Resistance

In the previous derivations, it was assumed that air resistance was negligible. This is generally true when fall distances are small. In such cases, the omission of air resistance factors produces a conservative solution of the vehicle's velocity, that is, the calculated velocity will be slightly less than the actual velocity. But from time to time it may be useful to estimate the air resistance in a vehicular fall, especially if the trajectory is long.

The drag due to air resistance is given by the following:

$$f_D = C\rho AV^2 \tag{xxvi}$$

where

f_D = air resistance force,
V = velocity through the air,
ρ = density of the air,
A = projected area of the vehicle normal to the direction of air speed, and
C = a constant which includes units conversion and the "slickness" of the vehicle.

For dry air at sea level and 0° Fahrenheit, its density is 1.293 g/l or 0.0807 lbs/ft^3. For variations in pressure or temperature, the application of the perfect gas law will suffice for four-place accuracy.

$$P/\rho = n\ R\ T \tag{xxvii}$$
$$P/(\rho T) = n\ R = \text{constant}$$
$$P_1/(\rho_1 T_1) = P_2/(\rho_2 T_2)$$

where

P = pressure,
T = temperature in absolute degrees,
ρ = density of the air,
n = number of moles of air, and
R = gas constant.

Dry air at sea level has a density 11% greater than sea level dry air at 30° Celsius. Corrections for temperature and pressure are sometimes significant, and should be considered. The relative humidity can also cause variations in the air density.

Inspection of equation (xxvi) finds that the drag varies with the square of the velocity directed into the air. Thus, at slow speeds the drag caused by air resistance is negligible. However, as the velocity increases, the drag quickly becomes more important.

One way to measure the frontal directed drag for a specific vehicle is to drive the vehicle along a level roadway when the air is still and note the loss of speed over time as the vehicle coasts to a stop (with engine disengaged). The following table lists actual data obtained for a 1983 Mazda GLC when allowed to slow down on a level, concrete roadway where there was no measurable wind.

Table 1. Slow Down Data - 1983 Mazda GLC

speed, mph	lapsed time, seconds
70	0
65	7.0
60	14.3
55	21.8
50	29.8
45	37.9

T = 100°F
radial tires
curb weight of vehicle and contents = 2,240 lbs.
frontal area = 3,000 in^2

Frictional forces in a free-rolling vehicle can be modeled by the following equation:

$$f = kW + C\rho AV^2 \tag{xxviii}$$

where

k = rolling resistance of the vehicle, and
W = weight of vehicle.

In general, the rolling resistance of a vehicle is basically constant for most vehicles and typical road speed ranges.

If equation (xxviii) is divided by the mass "m" of the vehicle, the retarding acceleration caused by the air resistance is obtained.

$$a = kg + C\rho AV^2/m \tag{xxix}$$

Using data from Table 1, the retarding acceleration can be calculated for several points using equation (xxix). For example, the acceleration between 70 mph and 65 mph is -1.052 ft/sec^2, and the acceleration between 55 mph and 50 mph is -0.913 ft/sec^2. Substituting these values back into equation (xxix) and solving two simultaneous equations gives the following:

$$kg = 0.7 \text{ ft/sec}^2$$

$$C\rho AV^2/m = 0.0000359 \text{ ft}^{-1}$$

Given that $g = 32.17$ ft/sec^2, then $k = 0.0218$.

Given that $A = 20.833$ ft^2, $\rho = 0.0663$ lbf/ft^3, and $W = 2{,}240$ lbf, then $C = 0.00181$ sec^2/ft.

The solved equation for the acceleration retardation by air resistance and rolling resistance is then:

$$a = -0.7 \text{ ft/sec}^2 - (0.0000359 \text{ ft}^{-1})V^2$$

If the Mazda were travelling at 100 ft/sec or 68 mph, the horizontal acceleration due to only air resistance would be -0.36 ft/sec^2. Therefore, over the velocity ranges normally encountered in falls, neglecting air resistance introduces little error.

Chapter 3: Example Problems

Problem 1

A car was found at the bottom of a ravine. Inspection showed that the ravine was 30 ft deep at the point of the car's impact, and the car was located 60 feet from the edge of the ravine next to the road.

At the top of the ravine, the ground was generally flat. Leading from the edge of the ravine were 25 feet of skid marks on hard ground where the car had apparently attempted to stop before going over the edge. The car weighed 2,400 lbs. How fast was the car going when skidding began?

Solution

Neglecting air friction, equation (vii) is applied to determine the time of fall of the vehicle.

$$t = [2(30 \text{ ft})(32.17 \text{ ft/sec}^2)]^{1/2} = 1.366 \text{ sec.}$$

Dividing the horizontal distance by the time gives the speed of the car when it passed over the edge of the ravine.

$$u = 60 \text{ ft}/1.366 \text{ sec} = 43.9 \text{ ft/sec} = 30 \text{ mph.}$$

The skid marks on the hard ground indicated that the following amount of work was dissipated:

$$U = (2{,}400 \text{ lbs})(0.4)(25 \text{ ft}) = 24{,}000 \text{ ft-lb.}$$

Applying equations found in chapter 2, section G, the velocity when skidding initiated is found to be 50.7 ft/sec, or 34.6 mph.

Problem 2

A stunt driver wishes to make a ramp-to-ramp jump of 75 feet using a ramp with a take off angle of 30 degrees. How fast should he go when he leaves the ramp in the direction of the trajectory? Compare a solution neglecting air resistance with one including air resistance.

Solution

Applying equation (xix) but letting "h=0", the following is obtained:

$$75 \text{ ft.} = 2(0.01346 \text{ sec}^2/\text{ft})V^2$$

$$V = 52.8 \text{ ft/sec or } 36 \text{ mph}$$

Assuming a frontal area of 23.5 ft^2, a dry air temperature of 80°F, a value of "C" of 0.002 sec^2/ft, and a vehicle weight of 2,800 lbs., then:

$$a_D = f_D/m = (0.000037 \text{ ft}^{-1})V^2$$

At 52.8 ft/sec, the deceleration due to air resistance is 0.103 ft/sec^2. The time necessary to cover the 75 ft. is then:

$$t = 75 \text{ ft}/(52.8 \text{ ft/sec})(\cos 30°) = 1.84 \text{ sec.}$$

Therefore, the velocity at the end of the jump would be about:

$$V_2 = 52.8 \text{ ft/sec} - (0.103 \text{ ft/sec}^2)(1.64 \text{ sec}) = 52.6 \text{ ft/sec.}$$

The reduction of speed by 0.2 ft/sec is a change of -0.3% from the solution where air resistance was neglected. Using the average speed of the car as it jumped from ramp to ramp, 52.7 ft/sec, the change in distance would be 74.8 ft. Thus, to adjust for air resistance, the jump distance should be reduced by 2.4 inches for a take off speed of 52.8 ft/sec (36 mph), or the speed should be increased to at least 52.9 ft/sec at take off to keep the distance at 75 ft.

As demonstrated in the above problem solution, air resistance can be considered insignificant for most purposes when the forward speed of a vehicle is 30 mph or less. Also, when air resistance is ignored, the determination of vehicle velocity will err on the low side.

Problem 3

If a car were dropped front first from an airplane, and it was able to stay oriented in that position through its fall, what would be its terminal velocity and how long would it take to reach it?

Solution

Assuming a frontal area of 3,400 in^2, a dry air temperature of 40°F, a vehicle weight of 2,600 lbs, and a "C" value of 0.0022 sec^2/ft, then the deceleration due to air resistance is as follows:

$$a_D = (23.61\ ft^2)(0.074\ lb/ft^3)(0.0022\ sec^2/ft)(V^2)/(80.82\ slugs)$$

$$a_D = (0.0000476\ ft^{-1})V^2$$

The acceleration of the car will initially be equal to "g", and will be diminished by drag until a constant velocity is attained.

$$g - a_D = a$$

$$g - (0.0000476\ ft^{-1})V^2 = dV/dt = 0$$

$$V = 822\ ft/sec \text{ or } 560\ mph$$

The time it takes to reach terminal velocity it given by the following differential equation:

$$g - KV^2 = dV/dt$$

where

$$K = 0.0000476\ ft^{-1}$$

The boundary conditions are: when "t=0", then "dV/dt = g"; and when "dV/dt=0", then "V=822 ft/sec".

A differential equation of this type is known as a *Riccati Equation*. It can be approximately solved in the time ranges in question by letting "V" be a truncated power series, substituting this power series back into the equation and then solving the unknown constants using the boundary conditions.

Let $V = A + Bt + Ct^2$

When "t = 0", "A = 0" and "B = g". When "dV/dt = 0" then:

$dV/dt = B + 2Ct = g + 2Ct = 0$

$t = -g/2C.$

Substituting the above into the non-zero remaining terms of the equation when "Dv/dt = 0" gives:

$t = 2/[g(0.0000476\ ft^{-1})]^{1/2} = 51.1$ sec.

Thus, the equation for velocity as a function of time in this situation is as follows:

$V = gt - (0.315\ ft/sec^3)t^2$

A table listing various values of the above equation follows.

Table of Equation Solution Values

t (sec)	V (ft/sec)	a (ft/sec^2)
0	0	32.17
5	153	31.06
10	290	28.16
20	517	19.43
30	682	10.06
40	783	3.00
50	821	0.09
51.1	822	0.00

Because the power series was truncated, it should be recognized that the solution equation fit is only between "t = 0" and "t" at the terminal velocity. Beyond "t = 51.1 sec", the solution is not meaningful.

CHAPTER 4

Acceleration and Deceleration

A. General

This chapter examines some of the basics concerning acceleration and deceleration of vehicles. One type of deceleration, skidding, has already been discussed in Chapter 2. This chapter is broader in scope, and discusses the theoretical underpinnings which affect vehicle response to acceleration and deceleration.

B. Engine Limitations

Each motor vehicle has an engine with a set maximum amount of power which it can deliver. Usually this is its horsepower rating. In the English unit system, a horsepower is defined as 550 ft-lbf/sec. In the metric or International System of units, power is usually measured in kilowatts, where 1 kw = 1.341 hp.

Power is defined as the amount of energy delivered per unit time.

$$P = d(E)/dt \qquad \text{(i)}$$

where

P = power,
E = energy, and
t = time.

In an engine, the amount of power delivered is normally a function of the engine's rpm's, revolutions per minute, and the torque exerted by the engine on the drive shaft before the clutch or transmission.

$$P = 2\pi T(rpm) \tag{ii}$$

where

T = shaft torque of the engine, and
rpm = revolutions per minute, a measurement of angular velocity.

As equation (ii) shows, for a given horsepower or power output, the torque increases as the engine rpm's decrease, and vice versa. Since the velocity of the vehicle is directly related to the engine rpm's, then it can also be said that the vehicle speed increases as the torque decreases.

As was noted in Chapter 3, the two main ways that energy is dissipated while a vehicle is traveling are tire friction with the road and air resistance. In Chapter 3, equation (xxviii) shows the frictional forces associated with rolling resistance and air resistance, and equation (xxix) shows these same effects as a deceleration. In general, rolling resistance is taken to include not only the specific tire and road rolling friction, but also the friction of bearings and other moving parts when the car is free-wheeling, i.e. when the engine is disengaged from the power train. For manual transmission cars, this is when the clutch is fully depressed. This is all lumped together into a single term designated as the rolling friction.

When the available engine power exceeds the energy dissipated by rolling resistance, air resistance, and any changes in elevation, the vehicle can accelerate. When the available power equals the dissipating effects, the vehicle will have reached its maximum speed (at those conditions). If any of the dissipating factors cause the dissipation rate to exceed the engine's available power, the vehicle will slow down until equilibrium is reestablished.

$$F_R = KW + CA\rho V^2 \tag{iii}$$

where

F_R = total frictional forces,
K = rolling resistance constant,
W = vehicle weight,
C = shape and units conversion constant (see Chap. 3),
A = front area of the vehicle,
ρ = density of the air, and

V = vehicle velocity.

It is assumed in the above that the wind velocity is zero, i.e., the air is calm. If this is not the case, the wind speed along the velocity direction of the vehicle would have to be added or subtracted accordingly.

Work is defined as the force time dot product of the distance over which it is applied. With respect to the energy dissipation factors, the work done by frictional resistance is as follows:

$$U = KW(x) + CA\rho V^2(x) \tag{iv}$$

where

x = the distance through which the force is applied.

If there is a change of elevation, then this must be taken into account by the factor "mgh", where "h" is the change in elevation. If the change in elevation is more or less linear, that is, "h = s(x)" where "s" is the change in elevation per horizontal distance, equation (iv) becomes the following:

$$U = KW(x) + CA\rho V^2(x) + mgs(x) \tag{v}$$

where

m = mass of the vehicle.

The dissipation of energy per unit time is therefore given by:

$$dU/dt = KWV + CA\rho V^3 + mgsV \tag{vi}$$

where "V" has been substituted for "dx/dt."

Setting the rated power of an engine equal to the right hand side of equation (vi) above, the maximum theoretical forward velocity of the vehicle can be determined.

$$P = V(KW + mgs) + CA\rho V^3 \tag{vii}$$

As was already noted in equation (ii), the engine power is a function of engine shaft torque and angular velocity. If this is considered equivalent to a force applied to the vehicle at a particular velocity, then "P = FV" where "F" is an equivalent force applied to the vehicle in order to move it, and "V" is the velocity it is traveling forward. Thus the force applied to the vehicle diminishes as its velocity increases and vice versa.

The above, of course, is known to most drivers from experience. More power is needed to pull out from a dead stop or go over a hill than is needed to maintain a cruising speed.

Using "P = FV" with equation (vii), gives the following expression for the force being applied to the car versus the dissipative fractional factors.

$$F = KW + CA\rho V^2 + mgs \qquad \text{(viii)}$$

The above equation allows calculation of the equivalent force being applied to the vehicle under the conditions at a steady velocity. It is similar to equation (xxviii) in Chapter 3.

Given the horsepower rating of an engine, equation (vii) can be used to determine the maximum forward velocity of the vehicle, provided the other factors are known. The maximum forward velocity can then be substituted for "V" in equation (viii), and the force that the engine can apply to the vehicle at that velocity can be determined.

Since "P = dE/dt" and "dE" would be the change in kinetic energy of the vehicle, then the following holds:

$$P[\Delta t] = \tfrac{1}{2} mV^2 \qquad \text{(ix)}$$

where

- P = engine horsepower rating,
- Δt = time interval,
- V = velocity, and
- m = mass of vehicle.

The above equation assumes that the initial velocity when the power was applied was zero.

Because there is a maximum "V" as already discussed, and assuming that the power output of the engine is constant over the time interval, then the amount of time it takes for the vehicle to reach the maximum velocity is as follows:

$$t = \tfrac{1}{2} mV_{max}^{\,2}/P \qquad \text{(x)}$$

The over-all or average acceleration from zero to maximum speed for the vehicle is then the following:

$$a = V_{max}/t \qquad \text{(xi)}$$

where t is from the solution of equation (x).

Generalizing, if it is assumed that the vehicle starts from a dead stop, then the velocity of the vehicle after time "t" is as follows:

$$V = [2Pt/m]^{1/2} \qquad \text{(xii)}$$

The above equation assumes that "t" does not exceed the amount of time needed to reach the maximum velocity. A plot of "V" versus "t" finds that initially, "V" is zero, but climbs fast. "V" tapers off somewhat asymptotically as "t" increases until the curve is essentially flat at the vehicle's maximum velocity.

Similarly, the vehicle's acceleration can be calculated as follows:

$$a = [2P/m\,t]^{1/2} \qquad \text{(xiii)}$$

Again, it is assumed that "t" does not exceed the "t" for maximum velocity found in equation (x). A plot of "a" versus "t" shows that "a" starts out asymptotically high, and then curves downward. As "t" increases, the curve approaches asymptotically to the "t" axis. At the time when "V_{max}" is attained, "a" is essentially zero.

Equations (xii) and (xiii) neglect frictional factors, power loss through the transmission, gear changes, and a number of other practical effects. However, the two equations do provide a sense of the way in which velocity and acceleration are expected to vary with time, engine power, and vehicle mass.

To model "V" and "a" more accurately, the frictional effects which would slow up the vehicle have to be accounted for. This is done as follows:

$$t = [\tfrac{1}{2}\,mV^2 + \tfrac{1}{2}\,KWVt + \tfrac{1}{2}\,mgsVt + (1/6)CA\rho V^3t]/[P(V)] \qquad \text{(xiv)}$$

where

P(V) = the engine power output as a function of vehicle velocity (or gear position).

Equation (xiv) assumes that the initial velocity is zero, and that velocity increases continuously, and more or less linearly.

C. Deviations from Theoretical Model

In the preceding section, it was assumed that all the power from the engine was transmitted to the car with 100% efficiency. This does not actually happen. Two significant factors reduce the engine efficiency in transmitting power to move the vehicle: wheel slip and transmission efficiency.

A brief definition of wheel slip was given in Section M of Chapter 2. When there is full traction between the tire and the road, there is no slip. This "no slip" condition occurs when the tangential velocity of the tires, where they contact the road, equals the forward speed of the vehicle. When the tangential speed of the tire differs from the forward speed, slip occurs.

As before, slip is defined as:

$$\text{slip} = S = 1 - [v_t/v] \tag{xv}$$

where

v_t = tangential velocity of the tire at the point of contact with the road, and

v = forward velocity of the vehicle.

To account for slip during acceleration, the term "(1 - S)" should be used to modify the velocity and acceleration calculations. For hard accelerations without "peeling", a slip factor of 0.15 to 0.20 is appropriate.

The second, and perhaps more significant factor, which affects the mechanical efficiency of transmitting power from the engine is the transmission efficiency. In an automatic transmission, is it not uncommon for the mechanical efficiency between power input and power output to be as follows.

Table 1. Common Automatic Transmission Efficiencies

gear position	efficiency "e"
1st gear	0.1 to 0.3 at high end
2nd gear	0.3 to 0.5 at high end
3rd gear	0.5 to 0.8 at high end
4th gear (if equipped)	0.8 to 0.95 at high end

This means that the available power "P" to move the vehicle has to be multiplied by an appropriate transmission mechanical efficiency factor "e", when calculating values for "V" or "a" when using the equations derived in section B.

D. Example

A 1988 Ford Taurus has a curb weight of 3,155 pounds and a net engine power of 140 hp. If it accelerates to 30 mph with an automatic transmission, it will be in first gear up to 15 mph, and then will switch to second gear.

Using the equation which neglects frictional forces, the time it would take to go from 0 to 30 mph would be as follows:

$t = \frac{1}{2} mV^2/[(1 - S)Pe$

$m = 98$ lbf-sec^2/ft

$V = 44$ ft/sec

$P = 77{,}000$ lbf-ft/sec

$S = 0.10$ (slip factor)

$e = 0.35$ (mechanical efficiency)

Substituting and then solving,

$t = 3.9$ sec and $a = 11.3$ ft/sec^2

For a speed of 60 mph, a slip of 10%, and average transmission efficiency of 60%, the time and average acceleration are calculated to be:

$t = 9.1$ sec $\quad a = 9.7$ ft/sec^2

From actual test data published by Consumer Reports (June 1988, pg. 399), the time needed to accelerate to 30 mph was 4.1 seconds, and the time for acceleration to 60 mph was 10.3 seconds. The respective accelerations were 10.7 and 8.5 ft/sec^2.

Not surprisingly, the approximation formula did well for the 30 mph regime, where wind resistance is negligible, but was less accurate in the regime above 30 mph, where wind resistance is more significant. Still, the correlation of the predicted values with the actual test values is good.

The maximum speed of the Taurus is estimated as follows:

Let

$A = 21.0$ ft^2

K = 0.03
C = 0.002 sec^2/ft
r = 0.075 lbf/ft^3
S = 0.05
e = 0.80 and
s = 0 (no elevation change).

Solving with the above values, we find that "V_{max}" for the Taurus is 227 ft/sec or 155 mph.

The air drag at 155 mph is calculated to be 1.65 ft/sec^2, and the rolling friction is calculated to be 0.97 ft/sec^2. The total calculated frictional deceleration is then 2.62 ft/sec^2 at maximum speed. It is estimated that maximum speed would be achieved in about 43 seconds and in about 5,000 ft. Overall acceleration from start to maximum would be 5.3 ft/sec^2.

E. Braking

There are published data for most vehicles concerning their braking ability. For example, the 1988 Taurus discussed in the last section was found to be able to brake without locking wheels from 60 mph to zero in 180 ft.

Applying some algebra to the simple dynamic equations finds that

$$a = V^2/(2d) \tag{xvi}$$

where

a = acceleration,
V = the initial speed prior to braking, and
d = the braking distance.

In the Taurus example, the braking deceleration is calculated to be 21.51 ft/sec^2 or 0.668g. Assuming that the driver was professionally trained, then the brakes were likely kept at a slip of about 20%. Therefore, the actual coefficient of friction between the test pavement and the tires would have been as follows:

$$f = (a/g)/(1 - S) \tag{xvii}$$

$$f = 0.67/0.80 = 0.83$$

In Chapter 2, Section D, Table 2, a frictional coefficient of 0.85 was given for dry concrete with little or no motion. Inserting the speed of 88 ft/sec into equation (xiii) in Chapter 2, a coefficient of 0.70 is calculated. Thus, the difference between a locked-wheel skid and a controlled hard braking with maximum effect is very small. In this case, it is a difference of only 4.6%. In fact, under most experimental conditions, the difference is less than the data scatter that would be expected over several trials.

F. Stuck Accelerators

The previous sections have demonstrated some facts that are relevant in "stuck accelerator" cases. These cases usually involve people who claim that their car or truck suddenly took off on its own, due to a stuck accelerator. In some of these cases, people will report that they tried to stop the car by pushing the brakes with all their strength, but that the car just kept on going.

With the 1988 Taurus, the initial acceleration was shown to be 10.7 ft/sec^2. This, of course, was by a professional driver who knows how to milk all the speed possible from a car. It was also shown that locked up brakes will generally cause deceleration at about 0.7g to 0.8g, i.e., 22.5 to 25.7 ft/sec^2. Comparing these two figures shows that no matter how hard the engine might try to run away with the car, the brakes exert over twice the deceleration needed to counter the runaway engine.

In short, the only way a stuck accelerator can cause a car to run away while someone is standing on the brake is for both the accelerator and the brake system to be dysfunctional at the same time. The subject of stuck accelerators and sudden acceleration incidents (SAI's) is discussed in more detail in Chapter 8.

G. Maximum Climb

Since a typical mid-sized passenger car has a take off acceleration of about 0.33g, this places a limit on the maximum grade that the car can climb.

When a box is laid on a frictionless inclined plane, its acceleration down the grade is as follows:

$$a = g \sin \alpha \tag{xviii}$$

where

α = the angle that the inclined plane makes with the horizontal.

In order for a vehicle to be able to go up an incline, the vehicle must be able to accelerate at more than the downward acceleration along the incline, and must be able to grip the roadway.

The perpendicular force of the vehicle pressing down upon the inclined plane is given by the following:

$$F = W \cos \alpha \qquad \text{(xix)}$$

where

W = weight of the vehicle.

In order for the vehicle to not simply slide down the incline, the friction between the tires and the surface must be greater than the sliding force parallel to the face of the plane.

$$Ff = Wf \cos \alpha > W \sin \alpha \qquad \text{(xx)}$$

where

F = force applied normal to the inclined plane.

$$f > \tan \alpha$$

If the coefficient of friction were 0.7, then the angle of the incline would have to be less than 35 degrees for the truck to not slide off with locked brakes on all four wheels.

In order for the vehicle to be able to climb the incline, the sliding acceleration must be less than the acceleration the vehicle can generate.

$$a > g \sin \alpha \qquad \text{(xxi)}$$

where

a = acceleration of vehicle, and

α = angle of incline.

If “a” is typically about 0.33g for maximum initial acceleration, then “a” must be less than about 19 degrees for the vehicle to make any headway up the hill.

It is noteworthy that in this case, the angle at which the vehicle stalls is much less than the angle at which it would lose traction with all four brakes locked.

There is still another item to consider in this situation: if the vehicle has four-wheel drive, then the above estimate of angle holds. If the vehicle is either front-wheel drive or rear-wheel drive, then an additional modification is required to account for the uneven distribution of drive through the wheels.

In a front or rear-drive vehicle, traction is provided by only one set of wheels. The frictional resistance to slide when the brakes are released and the engine is engaged is as follows:

$$Wf(\cos\alpha)\,x > W(\sin\alpha) \qquad \text{(xxii)}$$

where

x = fraction of vehicle weight carried by the drive axle.

$$f > (1/x)(\tan\alpha)$$

For front-wheel drive carrying 65% of the vehicle load with a coefficient of 0.7, the maximum angle is 24 degrees. For rear wheel drive carrying 35% of the vehicle load with the same coefficient of friction, the maximum angle is 13.8 degrees. This simple calculation demonstrates one of the advantages of front wheel drive.

It was shown that a vehicle which could produce 0.33g of forward acceleration could not make any headway up a slope greater than 19 degrees due to the limitations of power. The above has shown that if the vehicle has rear-wheel drive with a typical loading of 35% of the total weight on the rear axle, it will not have enough traction to make headway unless the angle is less than 13.8 degrees. While it would have sufficient power, its back wheels would simply spin and the vehicle would slide down. A front-wheel drive vehicle, however, would make it.

This rear-wheel drive effect also applies to the parking brake which in a passenger car or light truck locks only the rear wheels. It is typical for passenger cars that the rear axle carries only 35% of the vehicle's weight. Thus, the resistance to slide would be only 35% of that which would be afforded when all four wheel brakes are engaged by the foot pedal. This can fool a driver into thinking that because the foot brakes can hold the car on an incline, so can the parking brakes.

With drum brakes, this problem is further exacerbated because the brakes may be directional: they may be designed to provide more braking effect in the forward direction than in the reverse direction. Thus, a parking brake which may hold a car when the car is pointed down the incline, may allow the car to roll when it is parked heading up a steep incline.

H. Estimating Transmission Efficiency

Section C discussed briefly the efficiency factors associated with getting power from the engine to the wheels. Some representative values for various gears in an automatic transmission were given. As was shown, the mechanical losses at low speeds can be very high, while at high speeds the losses are much reduced.

Obtaining exact test information concerning mechanical efficiency of the drive train is often difficult, and doing the tests can be expensive. However, the values can be obtained indirectly from other performance parameters that are readily available.

Each vehicle manufactured or marketed in the U.S. must show two E.P.A. mileage ratings: one for highway driving and one for city driving. The highway rating is for flat highway driving at highway speeds. The city rating is for moderate speeds associated with city stop-and-go type of driving. Also, various consumer testing agencies publish similar fuel consumption figures. The ratio of the two values can be a useful indicator for the ratio of transmission efficiencies at the speeds tested.

For example, in the previous example of the 1988 Ford Taurus, the E.P.A. mileage ratings are 21 mpg city, and 29 mpg highway. The Consumer Reports figures are 15 and 29 respectively. This indicates that Consumer Reports tests for city driving at a lower average city speed than the E.P.A. The ratio of the E.P.A.'s figures is 0.72, and of the Consumer Reports' figures is 0.52. Since at highway speeds a typical transmission will have a mechanical efficiency of about 0.80, then the E.P.A.'s figures indicate that at their city driving speeds, the transmission efficiency is 0.58.

Similarly, the Consumer Reports figures indicate an efficiency of 0.42 at their city driving speeds. The E.P.A.'s figures are closer to the values associated with the low end of third gear, while the figures from Consumer Reports are closer to the values for second gear. The combination of these figures provides the three points from which a transmission efficiency curve can be constructed.

To demonstrate the consistency of these figures, consider the figures for a 1988 Buick Century with automatic transmission. The E.P.A. figures are 20/27 and the Consumer Reports are 16/29. The E.P.A. ratio is 0.74 and the CR ratio is 0.55. Using the same assumption that highway speed has a mechanical efficiency of about 0.80, then the E.P.A. city driving efficiency is 0.59 and the CR city driving efficiency is 0.44.

Compared to automatic transmissions, manual transmissions are generally more efficient at the lower end, and about the same at the upper end, i.e., about 80–85% at highway speeds. For example, a 1989 Eagle Summit LX has an E.P.A. of 29/35 and a CR of 25/37. These are ratios of 0.83 and 0.68 respectively. Assuming a highway efficiency of 80%, the respective city efficiencies are 0.66 and 0.54.

Similarly, a 1989 Ford Escort LX with manual transmission has an E.P.A. of 27/36 and a CR of 23/38. The respective ratios are 0.75 and 0.60, and the respective transmission efficiencies are 0.60 and 0.48.

The Escort and the Eagle both have 5-speed manual transmissions, each weighs about 2,340 pounds, both have similar weight distribution between the axles, and both have the same acceleration values for 0–30 mph and 0–60 mph (4.0 and 13.1 seconds, respectively). The Escort, however, has a four-cylinder, 1.9-liter displacement engine while the Eagle has a four-cylinder 1.5-liter displacement engine. Thus, because the Eagle has a smaller engine but has the same acceleration performance, it would be expected to have better low-range mechanical efficiency. This is borne out by the efficiency estimates.

While it is not as good as test data, this method does provide reasonable estimates of transmission efficiencies.

I. Estimating Engine Thermal Efficiency

As a point of interest, an estimate of the thermal efficiency of the vehicle's engine can be made by comparing the energy input, vis-a-vis the engine fuel consumption rate, to the power output, as shown in equation (iii).

A pound of gasoline has about 20,750 BTU (higher heating value) and a pound of diesel fuel has about 19,300 BTU. On a per-gallon basis, this is 128,000 BTU and 144,000 BTU respectively.

One horsepower is equivalent to 2,545 BTU/hr. Since highway fuel consumption is usually figured for 55 mph, then 1 hour of operation would use 55 mph/(x mpg), where "x" is the unknown mileage rating.

For example, a 1989 Nissan Sentra was found to get 44 mpg at highway speeds on flat, expressway type roads. Its engine power rating is 90 hp. Converting the units, 44 mpg is equivalent to a consumption of 160,000 BTU/hr. If the Sentra has a frontal area of 20.5 ft^2, a "K" value of 0.01, a "C" value of 0.008 sec^2/ft, and a weight of 2,275 pounds, then the frictional resistance at 55 mph would be about 103 lbf. Over the course of one mile that would be 544,000 lbf-ft or 699 BTU. For one hour, that would be 38,400 BTU. The engine efficiency would then be output/input or 38,400/(0.80)(160,000) = 0.30 or 30%. (The 0.80 value is to account for transmission efficiency.)

The above calculated value is typical for an efficient Otto or air standard cycle spark ignition engine.

J. Peel Out

When a vehicle initially accelerates at more than the tractive ability of its tires, it will leave an obvious peel-out mark. This mark may often provide valuable information concerning what the vehicle was doing at a specific location. This can be especially useful in SAI cases (see Chapter 8), and certain intersection cases.

In the case of peel outs, the torque applied to the tire by the engine exceeds the torque applied by the frictional resistance on the tire. This is expressed below.

$$T > f\,r(1/x)W \qquad \text{(xxiii)}$$

where

- T = torque engine applies to tire,
- $1/x$ = fraction of vehicle weight carried by tire,
- r = radius from center of wheel to where wheel contacts ground and is depressed,
- W = weight of vehicle, and
- f = coefficient of friction.

Equation (xxiii) explains why it is easier to peel out with a rear wheel drive car than with a front wheel drive car. Since most cars have their weight distributed more towards the front, a higher torque is needed for a front wheel drive vehicle to cause peel out, all other factors being the same.

K. Lateral Tire Friction

When a car is braked or decelerated, it can skid or have slippage in the longitudinal direction, i.e., the direction of travel. However, it can also slide or slip laterally. Due to the orientation of the car when braking begins, due to the crown of the roadway, or due to side impacts, a vehicle can also slide sideways while skidding forward.

In general, when a vehicle has no longitudinal slip, the friction between the tire and the road across the lateral axis follows the same type of calculation as before. It is still the same tire, the same roadway, and the same amount of contact area. Some variations in the coefficient of friction may occur due to tread orientation or edge effects of the side of the tire, but essentially a tire responds laterally about the same as it does longitudinally.

On dry pavement, the longitudinal coefficient of friction is maximum when the longitudinal slip is 10–20%. It then diminishes more or less linearly until 100% slip occurs. When the pavement is wet, the maximum longitudinal friction occurs when longitudinal slip is 30–50%.

On the other hand, lateral friction is maximum when longitudinal slip is 5–10%, and then diminishes rapidly when longitudinal slip exceeds 15%. At a longitudinal slip of about 30–40%, the lateral friction coefficient may be half of what it was when the longitudinal slip was 0%. Between 80–100% longitudinal slip, the lateral friction coefficient may be reduced to 0.10–0.20, with 0.15 being a typical value for both radial and bias tires when the pavement is wet.

L. Bootlegger's Turn

A "bootlegger's turn" is when a car is turned around 180 degrees, generally on a two-lane road, while still being driven forward. Generally, to get a car turned around, it is necessary to stop the car, back up and cut the wheels sharply, and then proceed forward. A bootlegger's turn simply spins the car around to face the opposite direction while it is still moving.

As the name implies, a bootlegger's turn is so named because it was a prohibition-era maneuver used by moonshine runners to escape pursuing "revenooers". Such quick turns would often catch the pursuers unawares, and they would overrun their quarry.

A bootlegger's turn is done as follows: while moving forward at a moderate or low speed, the emergency brake is suddenly applied so as to cause the rear wheels to slide. Just as the rear wheels begin to slide, the front wheels are steered sharply to the left or right. This will cause the rear end of the car to slide around. When the rear end of the car has spun sufficiently around to where the car can take off in the opposite direction, the emergency brake is released and the accelerator is mashed down to make a quick getaway.

What makes this maneuver work is the fact that when the rear wheels begin to skid longitudinally, the lateral traction significantly drops to where side skidding is very easy. However, because the front wheels are still rolling (and not significantly slipping), they retain their side traction. The car rotates around the front wheels which are still able to hold to the road, while the rear wheels are more or less free to slide sideways.

If the rear wheels do not fully skid before the sharp turn is executed, there is a risk of overturning the car. This is because the lateral coefficient of friction will not have dropped off enough. Thus, instead of sliding around, the car might roll over. Also, if the front wheels are allowed to skid also, as might happen if the driver used the foot brakes, the car would simply skid forward and not spin around sufficiently. For these reasons, it is not recommended that amateur drivers attempt this maneuver.

A similar maneuver called "doughnuts" uses the same principle of lower side traction due to longitudinal slip. In this maneuver, the car proceeds forward at a moderate or slow speed, usually in snow or on a loose gravel area. The driver mashes the accelerator to the floor so that the wheels spin or "peel out". By keeping the front wheels sharply turned, the car will spin around in a circle pivoting about the front wheels making a "doughnut" figure. In this case, the longitudinal slip occurs due to the peel out rather than a skid. By letting off the accelerator and allowing the rear wheels to "catch up", and straightening out the front wheels, the car can be driven out of the maneuver at will. This also is obviously not a recommended maneuver for an inexperienced driver.

Both maneuvers have been discussed because variations of these actions sometimes occur in an accident, often unintentionally, to a panicked driver. During panic stops when longitudinal slip exceeds 50%, the concomitant reduction in lateral traction can result in loss of control of the vehicle, especially when rotational forces are present.

The loss of lateral traction resulting from longitudinal skidding is often a causative factor in:

- truck accidents where jackknifing occurs.
- car accidents where skidding precedes loss of steering control of the vehicle, especially at curves and corners.
- motorcycle accidents where the driver looses control and lays the motorcycle down sometime after skidding initiates.

The primary advantage of an anti-lock brake system is that when the longitudinal slippage is limited to 30% or less, the reduction in lateral traction is held within sufficient limits to generally maintain control of the vehicle. This allows the driver to still steer the vehicle while braking as hard as possible.

Chapter 4 Example Problems

Problem 1

A man at the barbershop states that his car can reach 60 mph in less than 6 seconds. His car weighs 2,800 pounds, and the rated engine horsepower is 120. How much of this is braggadocio?

Solution

Applying equation (ix) and modifying it to include a slip factor of 0.1 and an engine mechanical factor of 0.6, then:

$t = 9.46$ seconds to reach 60 mph

Since equation (ix) neglects air resistance and tire resistance, this is a conservative figure.

If the slip were zero, and the engine mechanical factor was 100%, the time required would be 5.1 seconds.

Since the barbershop denizen's claim requires unrealistic performances from tires and transmission, i.e., the absence of tire and air frictional resistance, the acceleration claim is an exaggeration.

Problem 2

A small pickup truck has to climb a 5% grade that is covered with packed snow ($f = 0.20$). The truck has 70% of its weight towards the front, and is a rear-wheel drive vehicle weighing 2,400 pounds. Can the truck climb the grade?

Solution

A 5% grade is an angle of 2.86 degrees. Since the coefficient of friction is 0.20 and the rear wheels carry only 30% of the vehicle's weight, then:

$0.20 > [1/(0.30)(0.05)$

$0.20 > 0.17$

Since the inequality was found to be true, then the pickup would be able to drive up the slick hill. However, since the margin between the two factors is small, it would be good to add some sand bags in the rear to make sure.

Problem 3

What torque is needed for a 3,000-pound car with 60% of its weight in the front to climb a 10% grade with a coefficient of friction of 0.45? The car has rear-wheel drive.

Solution

The force needed to just equal the downward sliding force caused by the incline is:

$$F = W(\sin \alpha)$$

The torque at the wheel is related to the force by the following:

$$T = F\ r$$

where

r = the distance from the center of the wheel to point of contact of the tire with the road.

Since there are two rear wheels opposing the downward slide, then:

$$T = W(\sin \alpha)r/2$$

where

T = the torque each rear wheel applies to the ground.

In this case, if "r" = 16 inches, then the torque needed on each rear wheel to keep the vehicle from rolling backwards is as follows:

$$T = (3{,}000 \text{ lbs})(\sin 5.7)(16 \text{ in})/2 = 200 \text{ lbf-ft}$$

The frictional force that the rear wheels would exert if the rear wheels were locked is as follows:

$$F = (0.40)(0.45)(3{,}000 \text{ lb})(\cos 5.7) = 537 \text{ lbf}$$

The torque on each rear wheel that would be necessary to produce this amount of force is then:

$$T = (1/2)(537 \text{ lbf})(16 \text{ in}) = 358 \text{ lbf-ft}$$

Therefore, the torque exerted through each rear wheel which is necessary to simply hold the car in place is 200 lbf-ft. To accelerate the car up the hill, more than 200 lbf-ft of torque is needed. However, if more than 358 lbf-ft of torque is applied through each rear wheel, the wheels will peel out.

CHAPTER 5

Some Aspects of Motorcycle Accidents

A. General

The previous three chapters generally started with basic scientific principles from which were developed corollaries and equations applicable to accident reconstruction. This approach is useful in determining the factual aspects of an accident, i.e., the "what".

However, the following chapter about motorcycle accidents is somewhat different. This chapter gives information and equations that explain the underlying causative factors, the "why" of an accident. In part, it also demonstrates that some accidents have a built-in tendency to occur.

In our practice at Dressler Consulting Engineers, Inc., we have noted for some time that a large fraction of accidents involving motorcycles share certain common characteristics, including the following:

- the accident involves a motorcycle and a car, often a large car or truck.
- the accident occurs on a straight section of roadway which is often a simple two-lane road. The car and motorcycle are headed towards each other.
- the driver of the car or truck is usually older than 40 years.
- the driver of the car or truck is making a left turn, and the motorcyclist is proceeding straight ahead.
- the accident occurs when the motorcycle impacts the left-turning car.
- the driver of the car often indicates that he or she saw no approaching motorcycle.

- There are often no visual obstructions to impede either driver's view of the other.

Because this pattern has repeated itself many times, we did a small study to determine the underlying causes of such an accident. The following is what we discovered.

B. Background Information

According to the 1990 Statistical Abstract of the United States, in 1987 there were about 181 million registered vehicles in the U.S. This includes cars, trucks, and busses. Of this total, 139 million were automobiles. Additionally, there were 5.1 million registered motorcycles. In terms of the total number of registered vehicles, motorcycles constituted about 2.7%, cars constituted 75%, and trucks constituted 22.6%.

On the average in 1987, a registered motorcycle was driven only 17.6% as far in a year as the average registered car, truck or bus. The total road miles of motorcycles were only 0.5% that of the total for cars, trucks and buses.

From these statistics, it might be expected that the number of accidents involving motorcycles would be comparatively low. Based upon the single factor of total miles driven, it might be expected that the share of accidents involving motorcycles would be perhaps 0.5%.

The total number of car accidents in 1985 was 27.7 million. Truck accident totals were 6.1 million, and the total for motorcycle accidents was 440 thousand. Of the total of 34.2 million vehicle accidents, car accidents were 81%, truck accidents were 17.8%, and motorcycle accidents were 1.3%.

The total number of deaths in 1986 resulting within 30 days of a vehicular accident was 46,100. Persons in vehicles accounted for 33,700, pedestrians accounted for 6,800, motorcyclists accounted for 4,600, and bicyclists accounted for 900. Of the deaths related to vehicular accidents, motorcyclists accounted for 10% of the total.

In short, the statistics indicate that on a per-mile basis,

- motorcyclists have 2.6 times more accidents, and
- motorcyclists suffer 20 times more fatalities than drivers of other types of vehicles.

The reason for the higher number of fatalities per road mile of motorcycles versus other vehicles is readily apparent. In almost all severe collisions, the motorcyclist is ejected from the motorcycle and dissipates his or her kinetic energy by impacting the ground or another object, often the other vehicle. In this respect, the per-mile fatality rate simply reflects the relative safety of the vehicle configuration once an accident has occurred.

However, the reason why motorcycles have more accidents per road mile driven is not so readily apparent. Some may argue that a motorcycle should be able to more easily avoid an accident due to its greater maneuverability. Some may argue that since fewer infirm persons drive motorcycles than cars, the average motorcyclist is likely to have faster reflexes, and would therefore react faster to accident situations.

As noted above, a large proportion of motorcycle accidents involve motorcycles which are driving straight ahead, colliding with cars making left turns of some sort. This may be a left turn into a driveway or into a cross street. In nearly all these cases, the driver of the car reports that he or she never saw the approaching motorcycle, or did not see it until it was too late.

Thus, the perception of the motorcycle and motorcyclist by drivers of the other vehicles appear to be a key factor in the higher accident rate of motorcycles.

C. Headlight Perception

When a driver observes an approaching vehicle at night, the most readily apparent visual cue is the angular separation of the two headlights located on either side of the front of the vehicle. Since the distance from the left to the right headlight is more or less standard, an experienced driver can estimate the distance between his car and the oncoming vehicle by the amount of angular separation between the headlights. In order to estimate the closing speed, the driver observes the oncoming vehicle for a few moments to note how quickly the angular separation increases.

For example, a 1975 Ford Mustang's headlights are 52 inches apart. At 500 feet, the angular separation between the two headlights is 0.497 degrees. At 250 feet, the angular separation is 0.993 degrees. At 125 feet, the angular separation is 1.985 degrees.

The angular separation is given by the following equation:

$$\alpha = \arctan(x/d) \qquad \text{(i)}$$

where

x = distance from the center of left headlight to the center of right headlight,

d = closing distance between vehicles, and

α = angular separation in degrees.

Due to the small angles involved, the angular separation is for all practical purposes a linear function as follows:

$$\alpha = 57.3(x/d) \qquad \text{(ii)}$$

Even as close as 50 feet, the error of the linear approximation given in equation (ii) is less than 1%.

Thus, when the headlights are twice as far apart as when first observed, the oncoming car has halved the closing distance. This linear relationship is easy, and appeals to common intuitive notions of speed and distance. For most purposes, the small variations of headlight spacing among car models is negligible.

However, a motorcycle does not have two headlights spaced apart; it typically has only one front headlight. The angular separation of headlights is not there to cue a driver of the distance and speed of an approaching motorcycle. The only cue available is the light intensity of the single headlight.

Light diminishes in intensity as the square of the distance. In other words, a light with an intensity of 24 foot-candles at 50 feet will have an intensity of 6 foot-candles at 100 feet. The relationship is expressed as follows:

$$I = k/d^2 \qquad \text{(iii)}$$

where

I = light intensity,
k = arbitrary constant, and
d = the closing distance.

In order for a driver to estimate the closing distance of an approaching single-headlight equipped motorcycle, he or she must be able to compare the light intensity of the headlight as it is observed with some memorized value. This is difficult. Some brands of headlights are brighter than others. The aim of the headlight will affect the perceived brightness, and the aim of the headlight will be affected by the slope of the road. Furthermore, road dirt on the headlight will affect its brightness as will atmospheric conditions such as fog, rain, snow, haze, etc.

In order to estimate the speed of an approaching motorcycle, the driver must compare the intensity of the headlight at one instant with another and apply the inverse-square rule. As compared to estimating the speed of an oncoming car, the driver must now use a different and less reliable visual cue to determine distance, and must apply a more difficult and less familiar mathematical rule to estimate speed.

As was noted above, only 2.7% of registered vehicles in 1987 were motorcycles, and motorcycles constituted only 0.5% of all road miles driven. Most drivers are very experienced in observing two-headlight vehicles at night. However, motorcycles are statistically less frequently encountered. In cold climates, motorcycles may not even be encountered at all during the winter months.

Thus, as compared to vehicles with two headlights, a driver is more likely to misjudge the closing distance and speed of a motorcycle at night because:

- The visual cue is completely different
- The rule which applies to the cue to estimate speed is different
- The experience factor in applying the above two items is low.

D. Daylight Perception

The width of many cars is about 72 inches. The average width of a motorcycle without extensive fairing is about 30 inches; the width of a car is about 2.4 times that.

A person with 20/20 vision is able to readily distinguish and read 4-point type at a reading distance of 12 inches. A 4-point type letter is 4/72ths inches tall. Therefore, a person with 20/20 vision can distinguish and read type that subtends an arc of about 0.25 degree, which is about half the width of a full moon.

A person with 20/50 vision can readily distinguish and read materials at 20 feet that a person with 20/20 vision can distinguish and read at 50 feet. Some states allow persons with vision corrected to 20/50 to drive.

A person with 20/20 vision will fully recognize a vehicle during daylight hours when it subtends a horizontal arc of at least 0.25 degree. For a car, this occurs when the car is 1,375 feet away, or about a quarter mile. For a motorcycle, this will occur when the motorcycle is 573 feet away.

A person with 20/50 vision would not be able to distinguish the same car until it subtended a horizontal arc of 5/8ths degree. This would occur when the car is 550 feet away, or when a motorcycle is 229 feet away.

Figure 1. Can you see two motorcycles, one behind the other? They are both closer than the car.

Most two-lane highways are posted for no more than 55 mph, or 80.7 ft/sec. If two vehicles approach one another at the maximum legal speed, the closing velocity is 110 mph or 161.4 ft/sec. A driver with 20/50 vision would discern an oncoming car 3.4 seconds before they meet, but would discern a motorcycle only 1.4 seconds before they meet.

Generally, younger drivers have faster reaction times than older drivers. A reaction time of 1/2 second for young drivers and 1 second or more for older drivers is typical. If both an oncoming motorcycle and a car are traveling at 55 mph, an older car driver with 20/50 vision may barely have enough time for recognition and reaction before the two vehicles meet.

As has been noted, a significant portion of car and motorcycle accidents involve an oncoming motorcycle and a left-turning car. Since a modest car is about 16 feet long, a left turn from a stop will take about 19 feet of arc for the front end of the car to reach the left shoulder and another 16 feet further to get the car clear of the roadway. A moderate acceleration for a medium car from stop is 5 ft/sec^2. A simple calculation shows that the time required for a modest car to fully execute a left turn and be clear of the left lane is about 3.74 seconds.

If the driver of the car has 20/50 vision, he or she will discern a car at a distance of 550 feet away, and a motorcycle at 229 feet. This means that an oncoming car can be approaching at speed of 100 mph or less, and will be seen by the driver in time to avoid making a left turn in front of it. However, an oncoming motorcycle could collide with the left-turning car if the motorcycle were exceeding 61 ft/sec or 42 mph. The driver may not see the motorcycle in time before making the left turn, and the motorcycle may close the distance before the car can fully execute the turn.

The problem is further exacerbated because during hard braking, motorcycles require more driving technique, especially while turning. In fact, even under normal circumstances, a motorcycle typically takes longer to stop than a car. While the motorcyclist may see the left-turning car well in advance, he or she may not perceive that the driver is about to execute a left turn until it is too late to effect a safe stop.

A typical hard-braking deceleration for a motorcycle is about 16 ft/sec^2. At 42 mph, a motorcycle will need about 2.76 seconds to come to a complete stop; this does not include any reaction time. If the motorcyclist travelling at 42 mph does not perceive within one second that the car is initiating a left turn in front of his or her car, he or she will not be able to fully stop the motorcycle before impacting the car. The motorcyclist must then make an evasive turn of some sort to avoid making contact with the car.

Thus, while the above kinematic analysis indicates that car drivers with 20/20 vision will likely be able to discern a motorcycle well in advance of any problem, a car driver with 20/50 vision may not discern an approaching motorcycle in time. It may be no coincidence that the ratio of average car width to average motorcycle width is about the same as the ratio of motorcycle accidents per mile to other vehicle accidents per mile.

E. Review of the Common Factors

In Section A of this chapter, several items were listed as being common to this type of motorcycle and car accident. At this point all of them have been explicitly discussed except two: the age of the driver of the car, and the size of the car.

The larger size cars or trucks take longer to get fully across the roadway when making a left turn. The longer length needs just a little more time to clear the roadway, and takes up more broadside space across the roadway, making it a larger obstacle for the motorcyclist to avoid. Thus, the motorcyclist is more likely to collide with a long vehicle than with a short one.

With respect to the age of the driver, it is known that some persons experience a significant drop in visual acuity in their mid-forties or so. Reflexes also slow down in middle age. Up until the mid-forties, a person's visual acuity may be very stable, changing very little for many years. There would be little need for a change in glasses, or perhaps even a need for glasses. Often, persons who experience such a loss in vision are slow to recognize the problem.

Some states require that driver's licenses be renewed every three years. Often, the license renewal requires a vision test. A driver who has had a significant drop in visual acuity will likely have the problem detected at that time. If the drop in visual acuity occurs just after a license renewal, the driver could go as long as three years before having to correct the problem.

However, it was demonstrated that the failure to recognize the approaching motorcycle in time occurred when the car driver's vision was 20/50 and the speed of the motorcycle exceeded 42 mph. In many states, vision corrected to 20/50 is legal. Thus, the driver of the car may have sufficient visual acuity to satisfy the legal requirements, but insufficient to avoid the accident.

F. Difficulty of Finding a Solution

It might be argued that something should be done to correct this problem and save lives. However, the obvious solutions are not easy to deal with. Should people with 20/50 vision not be allowed to drive, or make left turns? Should motorcycles be outlawed, or perhaps forced to have two headlights spaced apart like a car? Should speed limits for motorcycles be lowered below 42 mph in areas where left turns by cars occur frequently? None of these "solutions" is politically acceptable, or even practical.

It has taken a long legal struggle just to get motorcyclists in most states to wear helmets, much less alter the basic design of their vehicles. And how may registered voters and car drivers are middle age and older? How well would they appreciate a law that would not allow them to drive unless they have vision correctable to 20/20? In short, the problem appears to be intractable. Fortunately, the number of persons injured in this type of accident is relatively small compared to the number of deaths and injuries from other types of accidents.

CHAPTER 6

Photographic Documentation

A. General

An engineer is often asked to analyze a vehicular accident long after it occurred. In such cases, the vehicles may have been sold long ago for scrap, and the original accident scene may have been changed by a highway upgrade project. Because some or all of the actual evidence can no longer be examined, the accident analysis must be done from available documentation. Primarily, this will consist of the police accident report and the insurance adjuster's file information.

Unfortunately, police accident reports vary greatly in quality from place to place, due to different training and experience. Further, important details may have been overlooked due to the pressures of caring for the injured, dealing with traffic flow, and removing debris from the roadway. Some important details are also often overlooked because, at the time of the accident, no one would have guessed that they would become an important issue.

For these reasons, any initial photographic documentation of the accident is an important resource for reconstructing an accident. If there is litigation several years later, a few well chosen photographs can make an important difference in the outcome of the case.

This chapter discusses some of the ways to quickly and thoroughly document accident information with photographs, and ways in which photographs of old accidents can be analyzed to provide information about items which are no longer physically available to examine.

B. Equipment

It is recommended that photographs be taken with a 35-mm single-lens reflex camera using 200 or 400 ASA speed color film. A reflex lens camera lets the photographer see what the shot will look like before it is taken, as the photographer actually looks through the same lens through which the photographs will be taken.

Good 35-mm photographs can be enlarged in size, made into slides, or examined under magnification to enhance detail. The negative allows many copies to be made, and sometimes development tricks can be used to highlight or bring out detail in the prints that may have not been otherwise apparent. With 400 ASA film, often it is not necessary to use a flash for indoor photographs or in dark places, like in a car trunk or under the chassis. A 200 ASA film is a general-purpose film for outdoor or well-lighted indoor areas. Both film speeds provide good detail resolution.

Generally, the use of black and white film is no longer recommended except for special purposes. Color has very useful visual characteristics for differentiating an oil stain from a water spot, a rust spot from dirt, a grease streak from a mud streak, and so on. If necessary, black and white prints can be made from the color negatives.

There are some special purposes, however, for which black and white prints are very useful. These applications occur when it is necessary to block out color blending which may be obscuring a detail outline. A color photograph of an item which "blends" into a similarly colored background may be easier to see in a black and white photograph which shows only tones of gray. For example, a camouflaged tank may not be easily seen in a color photograph, but may be easier to pick out in a black and white print. Camouflage uses color blending to obscure the outline of the tank against the background. A black and white print does not show the color blending as much.

The use of instant "Polaroid™" type film and cameras is generally not recommended for good evidentiary documentation. The photographs tend to have poor detail resolution. They are difficult to blow up or make into slides, and when they are, they generally appear "fuzzy." Sometimes the colors fade after several years. Since development occurs on the spot and there are no useful negatives, there is little opportunity to lighten or darken prints to bring out detail. Getting reprints can also be a problem.

Unfortunately, many insurance companies and police departments supply their employees with such instant type cameras and film. It is convenient to immediately staple an "instant" photograph of the damaged car into a file or report. However, the proliferation of one-hour photograph development shops has made even this advantage questionable.

C. Photographs

In using photographs to document information, it is important to recognize two rules:

- Film is much cheaper than trips. When in doubt, take a picture of the item.
- Take photographs as if you were a draftsperson preparing drawings of the accident particulars.

When a good set of photographs is taken, it may be unnecessary to even take any written notes as all the accident information will be in the photographs. While the accuracy of a person's written notes may be called into question, a clear photograph of the same detail is difficult to dispute.

D. Some Items to Photograph

Photograph all four sides of the vehicle, and if possible a top view. Preferably, the four views should be similar to an orthographic depiction of the vehicle. Similarly, like special features in a shop drawing that bear noting on a larger scale, special accident details should be photographed close up. This is done by using a macro lens, which can be attached to (or replaces) the normal lens of the camera.

When photographs are taken to mimic orthographic drawings of the vehicle, it is relatively easy to factor dimensions of unmeasured items in the photograph. This is done by comparing the unmeasured item to one whose dimensions are known. For example, it might be important to know to what depth the front end was crushed. Analysis of the top, right side, and left side views could provide that dimension even if the crush measurements had not been taken. This would be done by using, say, the known diameter of the tires in the photograph to provide a scale for the crush depth.

The tires should be photographed to show tread depth, wear patterns, and general condition. It is important to ensure that the tires observed on the vehicle are the actual ones used on the vehicle at the time of the accident. Sometimes replacement tires and wheels will be put on the vehicle to facilitate towing when the original ones are flat or damaged.

The insides of the wheels should be photographed. Brake cylinder failures normally result in a "wet" spot on the inside of the wheel. Similarly, the level of brake fluid may be useful to photograph, as many types of brake failures involve a loss of brake fluid in the reservoir.

It is useful to photograph the dashboard and controls to show their positions. It may be important to know whether the radio and air conditioner were on and the windows up when the car approached the railroad crossing or intersection. Under such conditions, the driver may not have been able to hear a train whistle or the air horn of a truck.

Similarly, it may be important to know whether the lights were on, the wipers were on, or the cruise control was on at the time of the accident, or if anyone has changed their position or setting which has already been documented by testimony.

In stuck accelerator cases, it is important to document the position of the gear selector at the time of accident. Also, the relative locations of the gas pedal, the brake, and the clutch are important. Any garbage on the floor, or rolling up of the floor mat, can also have some importance in such a case.

The interior space normally occupied by the driver and passengers is important to photograph. The presence of spilled drinks or food, open maps, cassette tapes, and similar items may indicate that the driver was distracted. The presence of liquor containers or drug vials has obvious importance. The items that invade the driver's or passenger's space can sometimes be matched to injuries in order to determine who the driver was, or where a person was sitting when the accident occurred.

A photograph of the driver's door edge will document the manufacturer's specifications and vehicle identification number (VIN). Oil change and maintenance information is sometimes contained on paper tags affixed to the driver's door edge. Establishing maintenance records can often be an important issue when a specific component has failed.

Sometimes it is useful to photograph the seat belts. Often, crush damage will pin the seat belts in their position at the time of the accident; this can establish whether they were being worn at the time of the accident. In many states the use of seat belts is mandatory, and some insurance policy rates are predicated on the driver promising to use them regularly.

Close up photographs of the filaments of the headlights or signal lights can often be used to determine whether the lights were on at the time of impact. Filament photographs can sometimes be used to determine if a turn signal or brake light was on when impact occurred. A severe impact will cause a hot filament to distend when the globe is unbroken. A broken globe will sometimes allow a hot filament to oxidize with the air, thus proving the light was on at the time of impact.

Similarly, a photograph of the fuse block helps to establish the condition of the circuits at the time of the impact. In most cases, a blown fuse can be seen in a photograph. A fuse that is good after an impact must have been also good before the impact. (Nothing comes to mind that is actually improved by a car accident.) Missing fuses might indicate that it was not possible for some components to have been operating at the time of the accident.

"Tag" marks are small patches of paint or other material which have been rubbed off one vehicle or item onto another by contact. Such marks should be photographed as they can provide easily understood evidence showing which vehicle hit which items, and in establishing the vehicle's trajectory during the accident. "Tag" marks can result from contact with other vehicles, signs, railings, barriers, trees, pedestrians, and similar.

E. Accident Scene Photographs

The approach to the point of impact by each vehicle involved in the accident should be photographically documented, preferably from the driver's viewpoint, being sure to include any warning signs, speed limits signs, and other sign information that would have been seen by the driver. Interestingly, a common mistake made in police accident reports is an incorrectly listed speed limit.

It is useful to make accident scene photographs at about the same time of day as when the accident occurred, so that the lighting and shadows are similar. For example, it might be important to know if the sun was directly behind the only traffic light at the time of the accident, or in the eyes of one of the drivers involved.

Any pavement markings resulting from the accident should be photographed. Radiator fluid spots, oil spots, transmission fluid spots, and other liquids can mark locations when certain things occurred during the accident. Because of the way close ups of pavement appear in a photograph, it is recommended that something be put in the photograph to provide a sense of scale or size reference.

F. Analysis of Photographs

The technique for analyzing accident photographs is basically reverse projective geometry. In projective geometry, a draftperson is given a set of dimensions for a part, and is asked to draw the part as it would appear from a particular location or perspective. With photographs, the perspective or view is given first, and the dimensions are to be determined.

A very simple example of this would be a photograph of a brick wall. To know how high the wall was, it is only necessary to count the number of brick courses in the photograph. Since a brick course has a standard thickness, the height of the wall would be the number of courses counted times the usual thickness of a brick course.

Suppose a car drove into a brick wall and crushed its front end to an unknown depth. Suppose also that the only evidence of the crush is a photograph of the right side of the vehicle after it was damaged. If the make and model of the car is known, then the original overall length of the car and the tire diameter are available from published references.

Within the photograph, the crushed length of the car can be measured in tire diameters. The length of the car after the accident would be its length in tire diameters times the tire diameter. The crush depth would then be the overall length of the undamaged vehicle less the scaled measured length of the damaged vehicle.

In this respect, a person does not have to painstakingly make a lot of measurements and notes to document the vehicle or accident scene. It is possible to substitute some of the detail survey work at the scene with appropriate documenting photographs which either show the information, or from which the information can be quickly scaled. For example, a good photograph of the VIN can save time as compared to copying it down. A good photograph of the side of the tire can save having to copy down all the tire information on the side. An overhead shot of the accident from a tall building or cherry picker can save having to tape out the basic dimensions of the intersection, and can be blown up later to make the scene sketch.

G. Videotape

A videotape of the evidence can also be very useful, as it gives the feeling of "being there" to those who watch it. A lot of evidentiary documentation can be done quickly with videotape which can be viewed later. However, videotape is not as easy to work with as flat photographs in scaling dimensions from the images.

When using videotape, a few simple rules should be observed:

- move the camera slowly from point to point, giving the viewer plenty of time to move with the camera.
- when the camera has to move to a new location, turn the camera off between the shots. Otherwise, a running camera shot done while the operator is moving appears like a thrill ride to the viewer.

- if scene descriptions are used in conjunction with the images, be cognizant of the fact that such descriptions must be factual and unbiased. Many courts will not allow the sound to be played during a videotape because it may be deemed to be biased commentary.

A videotape can be spliced and cut into a professional-appearing presentation of the evidence. Because most people in the U.S. watch more television than they read books or reports, jurors may be more "at home" viewing a well presented video tape than reviewing a report with photographs.

Videotape also has the advantage that it is done in "real time." Actions recorded on videotape can be timed from the tape. In fact, a video production shop can insert a clock right on the screen so that the viewer can see the lapsed time while actions are taking place. Some videocameras have an internal clock which places the date and time on the tape footage as it is shot. This is helpful in verifying the "real time" aspect of the tape.

On the other hand, videotape also has an insidious side to it. Clever editing and cutting can shorten or lengthen the time of a videotape to accomplish dishonest intentions. For example, some companies prepare simulations of accident events using computer generated images, or photographs that are computer manipulated. Time can be speeded up or slowed down as needed to make the proposed accident scenario "fit" the evidence. For example, a fall from a cliff can be simulated where the "g" being used is closer to 36 ft/sec^2 than 32.17 ft/sec^2 to reduce the fall time, and make the calculation for horizontal speed produce a higher number.

Just like in the cartoons, when simulated images are employed, the maker of the videotape can have the images do many small and subtle things which violate physical laws. Any accident scenario that can be imagined can be produced on videotape.

The problem is further exacerbated by the fact that sometimes it can be extremely difficult for the viewer to distinguish a "cartoon" from a "real time" videotape image. With computer imaging and manipulation, a simulated image on the videotape may not have the "look" of a cartoon. A simple example is color. A person may have green eyes in real life. However, using computer techniques, the person's eye color in a videotape can be changed to blue or black by substituting new color pixels in the same location as the old ones. When viewed on the videotape, the person appears natural. The change is not noted because it has been a change to something that the viewer can readily accept as normal.

Thus, a pink tag mark on a car can be changed to a white tag mark to implicate a different car in a multi-car pileup. Perhaps the car with the white color had a large liability insurance policy, while the car with the pink color is being driven uninsured.

With the proliferation of computer imaging programs and equipment, and the simultaneous proliferation of companies that do accident simulations on videotape, the problem of sorting out the contrived simulations from the "real" ones is likely to become even more difficult than it is now. When enough money is involved, there will be sufficient temptation for clever video image manipulation of the "evidence".

CHAPTER 7

Interpreting Lamp Filament Damage

A. General

Many vehicular accidents happen at night. In such instances, it is not uncommon for one party involved in the accident to claim that the headlights, taillights, running lights, brake lights, or turn signals of the other vehicle weren't on or did not work at the time of the accident. Of course, if the lights were in fact off at the time of the accident, there is a legitimate issue concerning the visibility of the unlit vehicle, and the prudence of the person driving it.

Even when the accident has not occurred during nighttime, there can be reasons for wanting to know if certain lights were on at the time. For example, in many cars when the car is in reverse gear, backup lights will come on automatically. If it can be shown that the backup lights were on at the time of the accident, then the position of the gear selector at the time of the accident can be verified.

Similarly, a common cause of motorcycle and car collisions is the failure of the motorist to see the smaller motorcycle. Newer model motorcycles are wired so that the headlight turns on whenever the motorcycle is operating. This is done in an attempt to improve the visibility of the motorcycle to motorists. If the headlight were to burn out a fuse, it might be possible to operate the motorcycle with the headlight not working. If it could be shown that the headlight of a motorcycle was not on at the time of the accident, then negligence on the part of the motorcyclist might be demonstrated.

Careful examination of the lamp filament and other particulars can often provide direct evidence that will indicate whether or not a lamp was on at the time of an accident.

B. Filaments

Commercial incandescent light filaments are nearly universally made of tungsten, a relatively heavy metal element (*W*) with an atomic number of 74 and an atomic weight of 183.85. At room temperature tungsten is a hard, brittle gray solid with a specific gravity of 19.3. As a comparison, iron has a specific gravity of 7.87. Tungsten is a transition metal and falls into the same periodic element family as chromium and molybdenum.

One of tungsten's attributes with respect to light filaments is that it has the highest melting point of all metals, 6,179°F. It is electrically conductive and at elevated temperatures can be drawn into wire. These characteristics make it an excellent material for incandescent light filaments.

In an incandescent light bulb, electric current is passed through the filament. The filament is a resistive load, and heats up due to the "I^2R" effect. When it has heated up enough, it emits visible light. Unlike many materials, tungsten has a melting temperature higher than the temperatures at which incandescence occurs, which ranges from 4,000° to 5,500°F.

C. Oxidation of Tungsten

Tungsten oxidizes in air at 752°F to form tungsten tri-oxide:

$$2W + 3O_2 \rightarrow 2WO_3 \tag{i}$$

As temperature rises, the rate of oxidation rapidly increases. Also, extremely fine tungsten is highly flammable and can ignite spontaneously in air.

Since the oxidation temperature of tungsten is well below the temperature at which incandescence occurs, to keep the tungsten from simply burning up, it must be encased in a bulb containing a non-reacting atmosphere such as nitrogen or carbon dioxide. While tungsten's reactiveness with oxygen may seem like a complication with respect to the manufacture of light bulbs, it is useful in determining whether a light was on at the time of an accident.

If the glass bulb around a lighted filament is broken open so that air can reach the filament, the filament will literally burn with yellow flame and white smoke. The resulting product, tungsten tri-oxide, does not dissolve in water. This is fortunate, because simple rainwater will not dissolve the tungsten tri-oxide away.

Sometimes the tungsten tri-oxide smoke residue can be observed on the headlight reflector as a whitish substance months after the accident, even if the vehicle has not been sheltered from the weather. However, tungsten tri-oxide will dissolve in caustic alkaline solutions.

Tungsten tri-oxide is normally canary yellow at room temperature. When heated, it becomes dark orange. It regains its original yellow coloring on recooling. Due to its yellow color, tungsten tri-oxide is sometimes used as a pigment in ceramics. Tungsten tri-oxide has a specific gravity of 7.16, and a melting temperature of 2,683°F.

If the tungsten wire has been exposed to air, but not sufficiently for extensive oxidation to take place, the filament will noticeably darken. This will be in contrast to the normally silvery, shiny appearance of the filament. The darkening is caused by light interference of a thin layer of tungsten tri-oxide on the surface.

Thus, if the bulb envelope was broken open during the accident, a person may reasonably conclude that the light was on at the time of the accident if:

- A yellow flame was observed at the light just after the accident.
- A yellowish or whitish powder or smoke stain is observed on the reflector, in the bulb remains, or elsewhere around the light.
- The filament wire is noticeably darkened.

Chemical analysis of the residual powder or smoke residue can further confirm that the material observed is tungsten tri-oxide. In this regard, the identification of tungsten tri-oxide by chemical or physical means is a relatively unique marker. There are few common uses for tungsten tri-oxide, except for coloring ceramics. Unless a person crashes his or her car into a ceramics shop carrying yellow pottery, the identification of tungsten tri-oxide near a broken light bulb is a very certain indication that the light was on when the bulb envelope was broken open.

In some cases it may be claimed that the lights were on when the envelope was broken open, but that the accident severed the wiring just before the bulb was broken open. Thus, the filament would not be affected. This argument is specious as the following demonstrates:

The oxidation temperature of tungsten, 752°F, is well below the temperature range of incandescence. When a light bulb is turned off, it takes some time for the filament to cool down below oxidation temperature. Since most accident events are measured in seconds or fractions of seconds, there is generally not enough time for a filament to cool down to avoid any oxidation effects.

Further, if the bulb is relatively cool with respect to the oxidation temperature, or little air reached the filament before it cooled, the filament coil will simply exhibit discoloration in proportion to its temperature when air reached it. Dark colors in the wire, such as black or purple, indicate a higher temperature than lighter colors like green or brown.

A rough indication of filament discoloration versus initial filament temperature is given in the following table.

Table 1. Filament Discoloration Versus Temperature

Observed color	Initial temperature of filament
green, brown	1,100°F
purple, black	1,200 – 1,300°F
yellow powder forms	+1,400°F
white powder forms	+4,000°F

Of course, if the bulb envelope is broken open, the total lack of any tungsten tri-oxide formation or coil darkening indicates that the filament temperature was below the oxidation point, and that the light had been off at least a short time prior to the accident.

D. Brittleness in Tungsten

At ambient temperatures, tungsten is very brittle; it exhibits no practical ductility. The yield point and the rupture point are practically the same. The material does not significantly stretch prior to breaking. This characteristic can be put to good use to determine if the lights were off, or at least relatively cold, at the time of the accident.

In accidents where the impact deceleration is high, the filaments may break apart. If they break apart with no discernible stretching of the filament coils, the filament was cold at the time of the impact, i.e., it was not on. Examination of the fractured filament ends under magnification would also show brittle type fracture, such as necking, cup and cone fracture, etc. with no significant indications of ductility.

E. Ductility in Tungsten

At elevated temperatures, tungsten becomes ductile. This characteristic allows it to be drawn into wires and filaments in the first place. The following table shows the relationship of temperature to the true plastic strain in tungsten.

Table 2. True Plastic Strain Versus Temperature - Tungsten

Temperature °F	Stress, psi	True Strain, in/in
600	65,000	0.1
800	65,000	0.2
1,000	65,000	0.3
1,200	65,000	0.4

True stress and true strain are the stresses and strains which occur in the material instantaneously, and are calculated using the actual cross sectional area rather than the nominal cross section area. True strain is generally calculated as follows:

$$\varepsilon = 2 \ln (d_o/d) \qquad \text{(ii)}$$

where

ε = true strain,
d = instantaneous diameter, and
d_o = nominal diameter.

The temperature at which tungsten changes from brittle to ductile is about 645°F. This is not too much less than the lower temperature bound for oxidation to occur.

If an impact occurs when the light is on and the bulb envelope is not broken open, it is still possible to confirm that the light was on due to ductile stretch of the filament coil, sometimes called "hot shock." Like a soft spring, the filament will simply stretch out in response to the impact. If the filament breaks, the fractured ends will exhibit necking down, reduced cross sectional area, or cup and cone fracture patterns.

By testing several light bulbs of the same make and model at various deceleration levels, it is possible to roughly calibrate the amount of coil stretch to the deceleration experienced by the bulb at impact. However, this is not to be taken directly as the deceleration experienced by the vehicle as a whole. The localized deceleration rate in the vicinity of the bulb can differ from the deceleration of the vehicle as a whole, and the former should not be taken as an automatic measure of the latter.

Similarly, the direction of distention of the filament coil is not an automatic indication of the direction of impact. The localized impact forces in the vicinity of the bulb can differ from those on the vehicle as a whole. Also, the filament coil can experience "rebound" of the coil, which would cause it to distend in the opposite direction. In cases where there are multiple impacts, the filament may have several superimposed distentions, making correlation to any one impact extremely tenuous.

Figure 1. Both filaments in tail light are distended.

However, if information is scant, some general indication of deceleration magnitude and impact direction can be learned from a careful study of the filaments as long as the above caveats are understood.

In cases where there are two filaments in a bulb, such as a headlight with both high and low beams, the lighted filament will indirectly heat up the unlighted filament. Thus, the lighted filament will stretch the most, or perhaps simply break. Such markers can be very useful in establishing whether the high beams or low beams were on at the time of the accident.

As a useful informational item, the low-beam filament in a sealed beam is usually the upper one, the one closest to the back dazzle shade. In the duplo type headlight, where the bulb is inserted through the back of the headlight, the upper filament is also the low beam. It is also nearest the back dazzle shade.

F. Turn Signals

Turn signals normally turn off and on about one or two times a second. While in a signalling mode, the cooling down time of a turn signal is no more than about 0.5 seconds. Oxidation of a turn signal light filament still occurs up to a second or more after the filament is turned off. Darkening and stretching can still take place several seconds after the turn signal is turned off. All of the markers used in steady-light type bulbs can be applied to turn signals and flashers to determine whether they were on at the time of the accident.

G. Other Applications

The filament stretching effect can be used to determine if any of the interior lights in a vehicle were on at the time of the accident, such as the dome light, the map light, etc. Of course, different make and model lights will respond differently to impact decelerations. While some will stretch readily, others may not discernibly distend under the same deceleration, and as inferred before, the local deceleration can significantly differ from point to point in the vehicle.

H. Melted Glass

When a glass bulb envelope breaks open, tiny shards of glass are produced. If some of these shards land on the hot filament, they may partially melt and stick to the filament. Glass typically melts at 1,200°F or more.

Examination of the filament under magnification is generally needed to observe this effect. Finding such melted glass particles on the filament can confirm that the filament was at least hotter than 1,200°F when the glass envelope was broken. Since incandescence occurs at temperatures well above 1,200°F, the presence of melted glass on the filament cannot automatically be taken to mean that the light was on; it only means that the light had been on just prior to the accident and had not yet cooled to less than 1,200°F.

I. Sources of Error

In cases where a vehicle has experienced a previous impact, the light filaments may exhibit stretch effects from the prior accident. In combination with new stretching, the combined direction or stretch may give an erroneous impact direction.

Repeated hard road shocks from pot holes can also cause filaments to stretch in other directions. Very old filaments can appear stretched due to normal high-temperature creep effects, sometimes called "age sag." Thus, the age and use of the vehicle should be noted when interpreting impact direction from stretched filaments. Age sag and slight distentions from hard pothole impacts could be misinterpreted as "hot shock" effects.

Some types of bulbs, usually taillight bulbs, will exhibit a darkened envelope as they age. This is due to the tungsten metal boiling out of the filament and being deposited on the inside of the envelope.

Pitting of the filament also indicates age. When a filament burns out due to age, the filament separates. At the point of separation, electrical arcing damage such as beading or necking down may be observed under magnification. Age sag or slight distentions from long wear can be distinguished from "hot shock" effects by the presence of the above secondary age characteristics.

Occasionally after an impact, an uninformed person will turn on the lights to see what still works. This can cause an otherwise unburned filament in a cracked bulb to burn up or darken. To check the circuitry after an accident, the use of a low-voltage ohmmeter is recommended rather than turning on equipment.

In relatively low-velocity impacts where the envelope is not broken, there may not have been sufficient deceleration to cause filament deformation or breakage. In such cases, the lack of filament markers can mean that the impact deceleration was just too low to come into play; it does not automatically mean that the lights were off. However, if the impact was too low to affect the filaments, this will be corroborated by correspondingly low damage levels.

Sometimes when the general impact is low, and the lights in question do not exhibit any "hot shock" effects, it is possible to observe some "hot shock" effects in other lights near the point of impact on the vehicle. If the lights which show "hot shock" effects normally turn on when the lights in question turn on, the operation of one set of lights can be inferred from the other.

CHAPTER 8

Sudden Acceleration Incidents

A. General

A sudden acceleration incident, or SAI, is defined as an unintended or unexpected vehicular acceleration. Sometimes SAI cases are informally called runaway car cases, or stuck accelerator cases. Typically, a person who experiences an SAI indicates that one or more of the following occurred:

1. The car takes off by itself from a stopped position, usually when the driver shifts from park to drive or reverse.
2. The car takes off by itself despite the driver's efforts to apply the brakes.
3. The car's cruise control appears to engage itself, causing the car to take off.
4. The driver reports that the car does not slow down when the accelerator is released.

The news media often give more attention to reported SAI's than to other accidents of the same severity. SAI's which are reported in the newspapers are often couched in language suggesting that the event was some sort of a personal betrayal or act of rebellion on the part of the car. Often, SAI's are described with phrases such as, "it fought the driver's best efforts to control it," or "it acted as if it had a mind of its own."

Perhaps this is due to the special relationship between people and their automobiles. People depend upon automobiles more than on any other machine in our modern society for their livelihood, recreation, socializing, and sense of personal freedom. Interestingly, several novels and motion pictures have explored this fear of car rebellion, most notably "*Christine*" by Stephen King.

B. Brakes versus the Engine

A car that can go from 0 to 60 mph in 8 seconds is considered to be a sporty vehicle. This computes to an average acceleration of 11 ft/sec^2. Some of us who drive more economical vehicles can expect 0–60 mph times of 10 to 14 seconds. These times represent average accelerations of 8.8 ft/sec^2 and 6.3 ft/sec^2, respectively.

In published acceleration tests from which these types of figures are obtained, the test driver literally floors the accelerator pedal and makes the vehicle go as fast as possible. Typically, the tests are done with factory-supplied, showroom type new cars, which are tuned and adjusted for peak performance. Such test figures represent the upper limits of acceleration of the vehicles using standard factory supplied equipment.

While the engine supplies the power to accelerate the car, the brakes consume power to decelerate it. Unless there is some mechanical deficiency in the brake system, most cars have a braking system capable of decelerating the vehicle about 22 ft/sec^2 or more. In a tug-of-war contest between the brakes and the engine, the brakes win. The brakes can decelerate the car two to four times faster than the engine can accelerate it. Even if only the emergency brakes are available (usually the two rear brakes), the deceleration effect of the emergency brakes alone is approximately equal to the acceleration which the engine can produce.

Consider the following example. According to the May 1989 issue of *Consumer Reports*, page 344, a 1989 Eagle Summit LX was found to accelerate 0–30 mph in 4 seconds, and 0–60 mph in 13.1 seconds. In braking, the same vehicle was found to stop in 150 feet from a speed of 60 mph without locking up the wheels.

The acceleration tests of 0–30 mph and 0–60 mph compute to an average acceleration rate of 11 ft/sec^2 and 6.71 ft/sec^2, respectively. (As shown in Chapter 4, the initial acceleration is higher than acceleration initiated at higher velocities. As the vehicle speeds up, less power is available from the engine to accelerate the vehicle.) The braking deceleration of the 1989 Eagle Summit LX without locking up the wheels computes to 26.0 ft/sec^2.

If a car is at rest and the brake pedal and accelerator pedal are mashed down simultaneously while the engine is in gear, the engine will stall, assuming that the clutch doesn't burn out first. Similarly, if the emergency brake is fully engaged and the engine is in gear when the accelerator is floored, the car will either not move and stall out, or it will lurch forward slowly, dragging along the locked rear wheels.

If the car is able to lurch forward, it is usually because the car has front-wheel drive (see Chapter 4, Sections E and G). Because the front of a car is typically heavier than the rear, deceleration by the locked rear tires alone may be slightly less than the acceleration transmitted through the unlocked front wheels of a powerful car. Therefore, some slight forward motion may occur. However, such motion would hardly be of a magnitude for consideration as an SAI, and could be easily controlled by the driver.

Thus, when a driver claims that the car accelerated into a "runaway" despite the driver's best efforts to apply the brakes, there are only a limited number of possibilities as to what occurred. They are as follows:

1. The brakes failed.
2. A pedal other than the brake was mashed down by the driver.
3. The driver has described the event "incorrectly".

If the car is still drivable after the accident, a simple drive around the block can verify point 1 above.

If sufficient information exists to quantify the acceleration which occurred in the SAI, such as a "peel out" mark or similar, then the acceleration magnitude of the runaway can be compared to the published 0–30 mph acceleration figures for the car model. If they are reasonably comparable, and the brakes are in working order, then it can be concluded that the driver simply mashed down on the accelerator instead of the brake as reported.

C. Power Brakes

In a car equipped with power brakes, the driver's foot effort in depressing the pedal is assisted by an auxiliary source of power. Usually this auxiliary power comes from one of the following:

- A pressurized hydraulic system powered by a pump operated off the engine.
- A vacuum diaphragm mechanism using low-pressure air sources from within the engine or from a vacuum pump run off the electrical system.

In either case, the auxiliary source of power allows the driver to use much less effort to engage the brakes than would otherwise be required. Power brakes are often used in large vehicles such a trucks, vans, recreational vehicles, buses and luxury cars where the driver's effort alone might not be enough to operate the brakes effectively.

If the auxiliary power system fails, the brakes will still function, but it will require much more effort for the driver to operate them. If a driver has become accustomed to power brakes and has never operated them without the power assist, the brakes may appear to not work. The brake pedal will be much harder to depress, and the resulting stopping effect may be reduced. If the driver is frail or weak, it may not be possible for the driver to operate the brakes.

In such cases, failure of the brake power assist could allow a driver to depress the brake pedal hard without noticeable braking response. The type of malfunction does mimic an SAI. If the power brakes malfunction as described, the engine drag can slow the car when the driver releases the accelerator pedal, and the emergency brake can be engaged for further stopping effect.

The loss of the brake power assist does not cause the engine throttle to open up and the car to suddenly accelerate. However, this type of malfunction may surprise or panic the driver when he discovers the lack of familiar response from the brakes. In such a panic, the driver may inadvertently step on the accelerator, thus causing an apparent SAI.

D. Linkage Problems

When a driver depresses the accelerator pedal, a mechanical linkage transmits the pedal motion to the engine throttle. Sometimes the linkage is constructed of wire cable, sometimes a series of connected rods, and sometimes a combination of the two. Other methods of transmitting the pedal motion to the throttle are possible (e.g. electromechanical, pneumatic, hydraulic, etc.), but a mechanical linkage gives the most reliable operation.

Most U.S.-built cars, and cars built to U.S. standards, have a spring load on the engine throttle such that if the linkage were to break or become disconnected, the throttle lever will return to the idle position. It does not return to an open or full throttle position.

In order to increase engine power output, the throttle must be opened. In order to do this, the throttle lever must be pulled against the opposing spring load. Whenever the force pulling on the throttle lever is relaxed, the throttle spring will return the throttle lever back to the idle position.

It is possible for the throttle spring to break or fail. In such cases, the accelerator might "hang" at the point where it was depressed furthest and not return to the idle position. If this were to occur, when the driver lets up on the accelerator, the car would continue to go at the same speed as before.

Similarly, a linkage can sometimes become bound or develop an unusual amount of friction, which could retard the return action of the spring. A bent rod, a "kinked" cable, or a wet cable that has frozen over can cause such a problem.

E. Cruise Control

Various types of control and sensing strategies are used in cruise control systems. Most cruise control systems utilize engine vacuum or a vacuum created by a small electric pump to control the throttle. Speed is sensed by a magnetic pickup on the drive shaft, by factory-installed sensors in the transmission, or by piggybacking off the signals of a digital speedometer, if the vehicle is so equipped.

Most cruise control systems have the following characteristics in common:

1. They have a set point for the speed which is to be maintained. The set point is usually engaged when the vehicle is traveling at the desired speed.
2. The set point is not absolute but has a tolerance above and below it before the throttle is incrementally opened or closed.
3. There is a minimum speed below which the cruise control will not operate. Most cannot be engaged when the car is in reverse or below 35 mph.
4. The set point can be increased or decreased without having to release the cruise control. The cruise control can therefore act like a hand throttle.
5. The cruise control is disengaged when the foot brake or clutch is tapped.
6. The cruise control can be re-engaged at the set point if the control lever is tapped, or the resume control is operated (same thing).

Cruise control SAI's are often related to some of the following situations:

1. The driver uses the accelerator to drive faster than the cruise control set point and forgets that the set point is still engaged. When the accelerator is released, the car does not fully slow

down, but slows down to the set point. The car appears to not respond to the released accelerator.

2. The driver engages the resume feature when travelling at a speed slower than the set point. Thus, the car appears to momentarily take off on its own.

3. The driver intends to tap the brake to disengage the cruise control but taps the accelerator instead. The car then appears to surge forward.

In all of the above situations, the cruise control has not malfunctioned.

Because most cruise control systems use vacuum (actually an air pressure slightly lower than the ambient) for controlling the throttle, the cruise control normally fails in a fail-safe mode. That is, when the cruise control fails and the vacuum is broken, the system ceases to function. A leak in any of the vacuum lines causes the system to stop, and a malfunction in any of the parts causes the system to stop. Also, when the brake pedal is depressed, this breaks the vacuum.

Cruise controls also use an integrated circuit "chip" to handle input and output signal information. While it is theoretically possible that such a chip could be affected by a nearby large electric or magnetic field, this has not been a practical concern. Induced false signal processing is thwarted by the fact that the internal redundancies in most cruise control systems require two or more simultaneous circuit failures to occur. This does not occur in practice.

F. Transmission Problems

In automatic transmissions, the linkage between the gear selector and the transmission can sometimes get out of adjustment so that the indicated gear on the selector is not the gear being engaged. Such out-of-adjustments can cause "half step" shifts in gear position. For example, the neutral gear might not be at the "N" position, but it might lie halfway between the "R" and "N" positions.

When such adjustment problems occur, they can cause a driver who sloppily shifts the gear selector to believe the car is in one gear setting when it is in the adjacent gear setting. This can cause unexpected movement of the vehicle and result in an SAI complaint.

Some out-of-adjustments develop slowly over time. At other times, they may be caused by recent mechanical work on the car which required temporary disconnection of the gear cable or linkage, and the reconnection work caused a shift in adjustment. Still other cases might involve worn detent pins or other such parts which allow slippage or "stop" in the gear position. In such reported SAI's, the gear selection adjustment problem will also be apparent after the accident; such problems do not spontaneously "heal."

Some late-model cars are equipped with automatic shift locks. This device requires that the driver put pressure on the brake pedal before the gear selector can be shifted out of "park." It is notable that since the introduction of this device, cars so equipped have had fewer reported SAI's. For example, during the period September 1986 through November 1988, Audis equipped with automatic shift locks had 60% fewer reported SAI's than identical model vehicles not so equipped. This result further corroborates the point that many reported SAI's are simply due to pedal misapplication.

G. Miscellaneous Problems

Some SAI's have resulted from floor mats, shoes, pop bottles and cans, debris or other items becoming entangled in the accelerator or brake pedal. Floor mats and shoes can sometimes roll under the brake pedal and prevent it from being depressed properly. Similarly, a folded-over floor mat or trash item may retard the return of the accelerator when released.

Some people slip off their shoes while driving and allow the shoes to lie on the floor in the foot well. This is especially true for women who wear high-heeled dress shoes. The shoes are often uncomfortable when worn for long periods of time, and can easily be slipped off. Stops and sharp turns can cause them to roll unnoticed under the car's foot pedals.

H. NHTSA Study

Because of the public interest in SAI's, the National Highway Traffic Safety Administration conducted a study of ten vehicle types with above-average SAI complaints. The results of the study, titled "An Examination of Sudden Acceleration" (U.S. Dept. of Transportation, DOT-HS-807-367) were made public in 1989. The following is a summary of the study's findings:

1. A substantial number of SAI's involved accelerator linkages which "hung up" due to a mechanical problem. However, the problem caused the throttle to remain at its last position; it did not cause further acceleration.

2. A few cruise control failures were noted. However, in all cases, application of the brake would cause the cruise control to disengage.

3. Extensive testing of cruise controls in strong electromagnetic fields found no SAI causation.

4. In thousands of SAI's reported to the NHTSA, brake failure occurred in only a handful of cases. Actual brake system failure plays no significant role in SAI's.

5. For most SAI's, the most plausible cause of an open throttle condition while attempting to brake was pedal misapplication, which was perceived by the driver as brake failure.

6. It was hypothesized that the vehicle models with higher frequency of SAI's had pedal placement or feel such that the driver could mistake one pedal for another, causing pedal misapplication.

7. There were reported incidents where drivers thought they were starting their car in "park" when it was actually in "neutral." When the driver acted to shift into "reverse", one notch back, the car would be in "drive." Often the selector indicator was broken or unreadable.

8. Recommendations to change pedal placement and design were made to enhance driver recognition of the pedal being applied.

CHAPTER 9

Momentum Methods

A. General

A milestone in the development of modern science was the publication of Newton's Three Laws of Motion. In brief, the three laws are as follows:

I. A body at rest tends to stay at rest and a body in motion tends to stay in motion, unless acted upon by an external force.

II. When a force is applied to a free body, the rate at which the momentum changes is proportional to the amount of force applied. The direction in the change of momentum caused by the force is that of the line of action of the force.

III. For every action by a force, there is an equal but opposite reaction.

The above three principles form the basis for all non-relativistic mechanics. Until vehicle velocities begin to reach about 3,000 km/s, about 1% the speed of light, the above three basic laws of motion will do for every instance of vehicular accident analysis. Since the present relative velocities of impacting vehicles rarely exceed 200 km/hr, or 0.0000185% of the speed of light, there is little need to apply Lorentz Transformations or other relativistic considerations to vehicular accident analysis.

Of course, this brings to mind the old story about the pedantic motorist who told the judge that he did not run a red light; he was traveling fast enough that the red light appeared green to him. The judge, no stranger to Doppler effect calculations himself, simply agreed with the motorist and fined him a dollar for every mile per hour over the speed limit.

In reviewing Newton's Three Laws, it is apparent that the analysis of momentum changes and forces is central to their application. In vehicular accidents, collision with a second vehicle usually causes changes in the direction and speed of both the involved vehicles. The analysis of the momentum and forces in such cases can provide valuable information about the accident, especially pre-impact speeds.

Sometimes when the available accident data is insufficient for other types of analyses to work, there is enough information to apply momentum methods to analyze the accident. Also, the application of momentum methods is an independent way of corroborating results from other types of analyses.

B. Basic Momentum Equations

The basic equation of engineering mechanics is the ever faithful:

$$F_A = m_A a_A \tag{i}$$

where

F_A = the applied force on item A,
m_A = the mass of item A, and
a_A = the resulting acceleration of item A.

Since the acceleration "a" is equal to the time derivative of the velocity, equation (i) can be rewritten as follows to more closely parallel Newton's Second Law.

$$F_A = (d/dt)(mv)_A \tag{ii}$$

where

v = velocity.

The term "(mv)" is referred to as the *momentum*.

By rearranging the terms in equation (ii), the following is obtained.

$$F_A(dt) = m_A(dv_A) \tag{iii}$$

In the above equation, it is assumed that mass is constant. Thus, the mass derivative term, "v(dm)," is discarded.

The indefinite integration of both sides of equation (iii), as done in equation (iv) below, produces a term called the *impulse*. If a constant force is applied to a body for a certain time, then the impulse on that body is simply the force multiplied by the time.

The amount of impulse applied to a free body is equal to the change of momentum which results from its application.

$$I_A = \int [F_A(dt)] = m_A[v_{A2} - v_{A1}] \qquad \text{(iv)}$$

Thus, a force applied on the body, "m_A", for a time duration "t" produces an impulse, "I_A." This impulse causes a change in the momentum of the body, which in this case equals the change in velocity vectors of the body.

Applying Newton's Third Law now allows a second body to be introduced into this analysis. If a second body is applying the force "F_A" to the first body, then the force on the second body will be equal but opposite in direction to that applied on the first body.

$$F_B = -F_A \qquad \text{(v)}$$

where

F_B = is the equal but opposite force applied on the second body.

Since the force applied on body A is equal and opposite to the force applied on body B, and since the forces on both bodies are applied for the same amount of time, then the impulse on body A is equal but opposite in direction to the impulse on body B.

$$-I_A = I_B \qquad \text{(vi)}$$

$$-\int [F_A(dt)] = \int [F_B(dt)]$$

$$m_A[v_{A1} - v_{A2}] = m_B[v_{B2} - v_{B1}]$$

By rearranging the above terms, the equation for the conservation of linear momentum for a two-body impact is obtained.

$$m_A v_{A1} + m_B v_{B1} = m_A v_{A2} + m_B v_{B2} \qquad \text{(vii)}$$

The left hand side of equation (vii) is the total linear momentum prior to impact, and the right hand side is the total linear momentum after impact. Equation (vii) is the fundamental equation used in momentum analysis of vehicular accidents.

It should be noted that all the terms involving velocity, acceleration, etc., in the above equations are vectors and should be treated accordingly.

C. Properties of an Elastic Collision

Properly defined, a fully elastic collision between two bodies is one in which the deformation of each body obeys Hooke's Law. Hooke's Law states that within certain limits, the strain of a material is directly proportional to the applied stress causing the strain.

However, a more practical definition of an elastic collision between two bodies is: a collision in which it is observed that the two bodies return to the same shape they had before the collision; they are not significantly deformed by the collision.

The second definition is more useful than the first because it can be generally determined by visual inspection. If a vehicle has sustained little or no permanent deformation in a collision, then the collision can be assumed to be elastic, barring any special impact energy absorption equipment. Importantly, if the second definition is applied and the collision is deemed elastic, then the first definition can be applied to analyze the impact.

In an elastic collision between two bodies, there are four stages to the collision:

1. Contact is made between the two bodies.
2. Deformation occurs on both bodies according to Hooke's Law, reaching some maximum.
3. Recoil or restitution occurs as the stored "spring" energy in each body is released.
4. Separation between the two bodies occurs.

If each body in the collision reacts like a linearly elastic spring, then the following Hookian rule is applied:

$$F = -ks \tag{viii}$$

where

k = spring constant,
s = amount of deformation of the body (a vector), and
F = the applied force.

According to Newton's Third Law, the force of each body acting on the other is equal in magnitude, but opposite in direction. Thus,

$$-F_A = F_B \qquad \text{(ix)}$$

$$k_A s_A = -k_B s_B$$

The ratio of deformation between the two bodies is then as follows:

$$s_A/s_B = -k_B/k_A \qquad \text{(x)}$$

Since the time duration in which the two bodies are in contact with one another is the same for both, then the following also holds.

$$-I_A/I_B = \{m_A v_{A1} - m_A v_{A2}\}/\{m_B v_{B2} - m_B v_{B1}\} = k_A s_A/-k_B s_B \qquad \text{(xi)}$$

$$[m_A v_{A1} - m_A v_{A2}]/k_A = s_A \qquad \text{(xii)}$$

$$[m_B v_{B2} - m_B v_{B1}]/k_B = s_B$$

A review of equations (x), (xi) and (xii) reveals two points. First, if the elastic constant "k" for both bodies is the same, then the resulting elastic deformation will also be the same for both bodies. Secondly, the elastic deformation of each body is a function of its own change of momentum and elastic constant.

D. Coefficient of Restitution

During a two-body collision, there will be a point where the deformation of each body is at a maximum. When this point is reached, the two bodies will also have the same velocity. This common velocity is calculated by the following, using the pre-impact momentum sum.

$$[m_A + m_B]u = m_A v_{A1} + m_B v_{B1} \qquad \text{(xiii)}$$

$$u = [m_A v_{A1} + m_B v_{B1}]/[m_A + m_B]$$

where u = the velocity at which both bodies are momentarily moving together.

This velocity can also be calculated from the momentum sum after the collision.

$$[m_A + m_B]u = m_A v_{A2} + m_B v_{B2} \qquad \text{(xiv)}$$

$$u = [m_A v_{A2} + m_B v_{B2}]/[m_A + m_B]$$

When the two bodies momentarily have the same velocity "u", the portion of the collision which occurs before it is considered the "deformation portion" of the collision, and the portion which occurs after it is considered the "restitution portion." Therefore, the impulse for the deformation portion of the collision can be calculated from the definition of impulse given in equation (iv).

$$I_{def} = m_A[v_{A1} - u] = m_B[u - v_{B1}] \tag{xv}$$

where

I_{def} = impulse during the deformation portion of the collision.

Likewise, the impulse for the restitution portion of the collision can be calculated from the definition of impulse given in equation (iv).

$$I_{res} = m_A[u - v_{A2}] = m_B[v_{B2} - u] \tag{xvi}$$

The ratio of the restitution impulse to the deformation impulse is called the *coefficient of restitution*, "e."

$$e = I_{res}/I_{def} = [v_{B2} - v_{A2}]/[v_{A1} - v_{B1}] \tag{xvii}$$

It should be noted that the collisions being considered are assumed to be central collisions. If a collision is oblique, then the component vectors have to be separated. Those component vectors that are involved in the central collision obey the same relations for the coefficient of restitution as noted above. Those components that are not involved in the central collision are handled separately. The latter vector components are often involved in causing rotation, or doing frictional work such as sideswiping.

E. Properties of a Plastic Collision

In a non-elastic or plastic collision between two bodies, there are also four basic stages of the collision:

1. Contact is made between the two bodies.
2. Deformation occurs on both bodies reaching some maximum amount.
3. Some recoil or restitution may occur as any stored "spring" energy in either body is released.
4. Separation may occur between the two bodies if recoil is sufficient.

The main differences between the four stages of a plastic collision versus an elastic one are as follows:

1. The deformation does not follow Hooke's Law, that is, it does not elastically deform. It is likely that some elastic deformation will initially occur, but it may be insignificant compared to the amount of plastic deformation that occurs.
2. There may not be any significant recoil or restitution.
3. The two bodies may not have sufficient recoil to separate; they may stay "stuck" to one another after impact and move as a unit.

A perfectly elastic collision has a coefficient of restitution of 1.0, and a perfectly plastic collision has a coefficient of restitution of zero. Between those two extremes are combinations of the two, where some elastic and some plastic deformation occurs. Obviously, the lower the magnitude of "e", the less elastic deformation there is with respect to the plastic deformation.

Usually, any elastic deformation which occurs in a collision takes place in the early part of contact between the two bodies. The plastic deformation then takes place after the elastic deformation range has been exhausted, and stops when the two bodies reach the "u" velocity (see equation xiii). Consequently, as the severity of the collision increases, the value of "e" approaches zero. Conversely, as the severity of the collision decreases, the "e" value increases. Generally, the severity of the collision is proportional to the relative closing speed of the two vehicles.

The following coefficients of restitution, or "e" values, have been found to occur in passenger vehicles either colliding with a fixed barrier or its equivalent. An example of a fixed barrier is a well-built brick wall which does not move or become damaged when it is impacted. Letting the subscripts "A" denote a vehicle and "B" denote a fixed barrier, then the coefficient of restitution of a fixed barrier is equal to the following:

$$e = [v_{B2} - v_{A2}]/[v_{A1} - v_{B1}] = -v_{A2}/v_{A1}$$

because $v_{B2} = v_{B1} = 0$

When forward or rearward velocities are 2.5 mph or less, "e" may be considered to be 0.9–1.0. This is because U.S. Federal Law, as embodied in the National Traffic and Motor Vehicle Safety Act of 1966 with amendments, requires that bumpers on cars made after 1977 be able to withstand a 2.5-mph front or rear impact with a fixed barrier without significant damage occurring to the vehicle (49 CFR 581, 1987 Edition).

For a time before 1978, vehicle bumpers were required by law to be able to withstand a 5-mph impact with a fixed barrier. However, this was rescinded in 1977, and the current, lower standard was established. Before 1966, there were no legally mandated bumper standards. Thus no similar "e" value can be categorically assigned to vehicle bumpers made before 1966.

From experience, it has been found that at impact speeds of about 25 mph, an "e" value of 0.2 is typical for most passenger cars. At impact speeds of about 35 mph, an "e" value of 0.1 is typical, and at speeds exceeding 50 mph, an "e" value of about 0.004 is typical.

Interestingly, when the above experience-derived values of "e" are plotted against the vehicle impact speed, they tend to follow closely the following equation:

$$e_{res} = e^{-0.065s} \qquad \text{(xviii)}$$

where

e_{res} = the coefficient of restitution,
s = the vehicle's speed in mph, and
e = natural logarithm base "e", i.e., e = 2.718

The subscript "res" was used to differentiate the coefficient of restitution "e" from the natural logarithmic base "e".

F. Analysis of Forces During a Fixed Barrier Impact

In experimental tests, it has been found that when a 1980 Chevy Citation weighing 3,130 lbs was driven into a fixed barrier at 48 mph, the front end crushed 40.4 inches deep across the front of the car.

Using equation (xviii), it is found that this impact velocity would have a coefficient of restitution of about 0.044. Using this "e" value, the calculated rebound velocity would have been about 2 mph. For all practical purposes, this impact would be deemed a plastic collision.

The pre-impact momentum of the car would have been:

$$[(3{,}130 \text{ lbf})/(32.17 \text{ ft/sec}^2)][(70.4 \text{ ft/sec})] = 6{,}850 \text{ lbf-sec}$$

The post-impact momentum of the car would have been:

$[3{,}130\ \text{lbf})/(32.17\ \text{ft/sec}^2)][3\ \text{ft/sec}] = 292\ \text{lbf-sec}$

The net change in momentum due to the impact was then 6,558 lbf-sec.

After contact with the wall, the car came to a stop in 40.4 inches. Assuming that the velocity of the vehicle decreases linearly (i.e., constant deceleration), then the equation for the speed of the car during impact would be as follows:

$$70.4\ \text{ft/sec} - (1.743\ \text{ft/in-sec})x = V$$

where

x = the amount of crush which has occurred, and
V = the velocity at that point during the impact.

If the average velocity during the impact is simply "(1/2)(70.4 ft/sec)", then the time required to stop in 40.4 inches would be 0.0956 seconds. Thus, the actual impact is literally over in less than the blink of an eye.

Assuming that the force applied by the wall to stop the car was constant, then the impulse during the impact would be:

$$F(0.0956\ \text{sec}) = 6{,}558\ \text{lbf-sec}.$$

$$F = 68{,}598\ \text{lbf}.$$

If the force had been applied more or less evenly across the front of the car, and the front is assumed to be about 72 inches wide and about 30 inches high, then an equivalent average pressure on the wall to resist the car impact would have been 32 psi.

The deceleration rate during the impact is estimated as follows:

$d = \frac{1}{2} at^2$
$d = 40.4$ in
$t = 0.0956$ sec
$a = 8{,}841\ \text{in/sec}^2 = 737\ \text{ft/sec}^2 = 22.3$ g's

From this simple analysis, a few items are noted. First, if the crush depth is increased, then the deceleration can be decreased and the time of impact increased. For example, if the crush depth had been 50 inches instead of 40.4 inches, then the time of impact would have been 0.118 seconds, and the deceleration would have been 18.6 g's.

The 9.6-inch increase in crush depth, a 24% increase, would have reduced the deceleration by 17%. This is significant because the degree of injury sustained in a vehicle impact by the occupants is a function of the impact deceleration.

H. Energy Losses and "e"

When two bodies collide, some of the total kinetic energy is dissipated in plastic deformation. The amount of energy loss is given by the following.

$$\tfrac{1}{2}(m_A v_{A1}^2) + \tfrac{1}{2}(m_B v_{B1}^2) = E_{Loss} + \tfrac{1}{2}(m_A v_{A2}^2) + \tfrac{1}{2}(m_B v_{B2}^2) \qquad \text{(xix)}$$

$$E_{Loss} = \tfrac{1}{2} m_A(v_{A1}^2 - v_{A2}^2) + \tfrac{1}{2} m_B(v_{B1}^2 - v_{B2}^2)$$

In the above equation, the "v^2" terms are considered either vector dot products, or simply the absolute values of the velocity vectors squared.

The coefficient of restitution "e" is given by the following:

$$e = I_{res}/I_{def} = [v_{B2} - v_A^2]/[v_{A1} - v_{B1}]$$

Rearranging the above relation for "e" to favor post-impact velocity terms gives the following:

$$v_{B2} = e[v_{A1} - v_{B1}] + v_{A2} \text{ and} \qquad \text{(xx)}$$

$$v_{A2} = v_{B2} - e[v_{A1} - v_{B1}]$$

Using the above value for "v_{A2}", the following is generated.

$$(v_{A1}^2 - v_{A2}^2) = v_{A1}^2 - [v_{B2}^2 - 2(e)(v_{A1} - v_{B1})(v_{B2}) + e^2(v_{A1} - v_{B1})^2] \qquad \text{(xxi)}$$

$$(v_{A1}^2 - v_{A2}^2) = v_{A1}^2(1-e^2) - v_{B2}^2 + v_{A1}v_{B2}(2e) - v_{B1}v_{B2}(2e) + v_{A1}v_{B1}(2e^2) - v_{B1}^2(e^2)$$

Similarly, using the value in (xx) for "v_{B2}", the following is generated.

$$(v_{B1}^2 - v_{B2}^2) =$$

$$v_{B1}^2(1-e^2) - v_{A2}^2 + v_{B1}v_{A2}(2e) - v_{A1}v_{A2}(2e) + v_{A1}v_{B1}(2e^2) - v_{A1}^2(e^2) \qquad \text{(xxii)}$$

Substituting the above velocity difference terms, equations (xxi) and (xxii), into equation (xix) gives the following for the energy loss:

$$E_{Loss} = \tfrac{1}{2}(m_A)[v_{A1}^2(1 - e^2) - v_{B2}^2 + v_{A1}v_{B2}(2e) - v_{B1}v_{B2}(2e)$$

$$+ v_{A1}v_{B1}(2e^2) - v_{B1}^2(e^2)] + \tfrac{1}{2}(m_B)[v_{B1}^2(1 - e^2) - v_{A2}^2 + v_{B1}v_{A2}(2e)$$

$$- v_{A1}v_{A2}(2e) + v_{A1}v_{B1}(2e^2) - v_{A1}^2(e^2)] \qquad \text{(xxiii)}$$

Rearranging equation (xxiii) to sort out the "e" terms, and substituting to favor post-impact velocities, gives the following expression:

$$E_{Loss} = [(1/e^2) - 1][(m_A m_B)/2(m_A + m_B)][v_{B2} - v_{A2}]^2 \qquad \text{(xxiv)}$$

Similarly, if pre-impact terms are favored, the following expression can be derived:

$$E_{Loss} = [(1 - e^2_{res})/2][(m_A m_B)/(m_A + m_B)][v_{A1} - v_{B1}]^2 \qquad \text{(xxv)}$$

When "e = 1", both equations (xxiv) and (xxv) become zero, indicating again that a fully elastic collision is one in which there is no energy loss. When "e = 0", equation (xxx) is indeterminate due to division by zero. However, equation (xxv) is solvable.

To have a half-plastic and half-elastic collision, "e" must equal 0.707. In a passenger car colliding with a fixed barrier, such a half--plastic and half-elastic collision would occur at a speed of about 5.33 mph. A collision with only 10% elastic energy would have an "e" of 0.316 and would occur at a speed of about 17.7 mph. It is readily apparent, then, that in most vehicular collisions the dominant deformation mode is plastic. It is also apparent that in most vehicular collisions, recoil will be a small effect.

I. Center of Gravity

In all of the foregoing analyses, it has been tacitly assumed that the masses of the bodies were point masses. They were considered to be like minute billiard balls, bouncing off one another. When such equations are applied to vehicles involved in collisions, they then apply to the centers of mass of the vehicles.

Thus, all of the previous equations were based upon the premise of straight central impact, where the line of action between the bodies was through the center of gravity of each body.

As was noted in Chapter 2, Section L, the distribution of load on a vehicle is not necessarily such that the center of mass is at the geometric center of the vehicle. If "Q" is the load carried by the rear axle and "R" the load carried by front axle, then the center of gravity in a two-axle vehicle will be found as follows:

$$a = Qb/(Q + R) = Qb/W = b/(1 + R/Q) \qquad \text{(xxvi)}$$

where

a = distance from the front axle to the center of gravity of the vehicle,
b = distance between axles, and
W = total weight of the vehicle.

For example, in a car with a 106-inch wheelbase and 60% of the curb weight carried by the front axle, the center of gravity is located 42.4 inches away from the front axle in a direction towards the rear of the car. Alternately, the center of gravity is located 13.6 inches away from the midpoint of the wheelbase towards the front of the car.

The above, of course, assumes that the center of gravity is found on a line that bisects the length of the vehicle. For most purposes, this will be sufficient. If, however, there is some type of asymmetrical load on the vehicle, perhaps a load shift during the collision, it might be necessary to determine the center of gravity using the point mass method. In this method, the weight carried under each tire is determined, and the coordinates of the center of gravity are determined by the following:

$$x_c = [\sum m_i x_i]/[\sum m_i] \qquad y_c = [\sum m_i y_i]/[\sum m_i] \qquad \text{(xxvii)}$$

where x_c and y_c are the coordinates of the center of gravity.

Typically, the point of origin from which the "x" and "y" distances of the mass points are measured is one of the tire contact points. This causes the "x" and "y" values for that tire to be zero, thus simplifying some of the calculations.

Sometimes it is important also to know the position of the center of gravity with respect to the "z" axis, that is, the vertical axis. However, while the center of gravity with respect to the horizontal plane and "x and y" coordinates is relatively easy to locate, the "z" coordinate requires more effort.

Perhaps the simplest, although not the easiest, method of determining the center of gravity with respect to all three coordinates utilizes a crane. The vehicle is lifted with a three-point or four-point harness and is suspended by a single cable attached to the harness. Preferably, the vehicle should be somewhat tilted or otherwise not parallel with the ground. While the vehicle is suspended, a photograph is taken of it from a set position. A second lift is then similarly done except the vehicle is tilted differently from the first lift. With the vehicle facing the same side towards the camera, a second photograph is taken.

The single cable that holds up the vehicle acts as a plumb line in the photographs. The line of action of the cable runs directly through the center of gravity of the vehicle. Thus, the line of this cable should be graphically extended so that it runs down through the image of the vehicle. When the two photographs are superimposed, the point where the two plumb lines intersect on the vehicle is the three-coordinate center of gravity. If the photographs were taken with the same side facing the camera, it is an easy matter to locate this point on the photographs and then determine where it is on the car itself.

Unfortunately, apparatus for lifting cars and taking their picture is not always handy, or cheap. Thus, an alternate method may need to be used which has a lesser equipment requirement.

Knowing that the "z" component of the center of gravity will be directly above the "x and y" components, then the "z" component can be found by jacking up the vehicle with a scale under the jack to read the load. The calculations are easiest when the jack is placed in line with either the longitudinal or transverse axes of the C.G. By jacking up the vehicle until the initial load on the jack is significantly reduced, and noting the angle of the vehicle, the "z" component can be found by simple statics.

If a person were to continue to jack the car up until it is balanced on the two wheels making contact, the center of gravity would be directly over the axis of the two wheels, and the load on the jack will be zero.

By taking a photograph with the view parallel to the axis of the contacting wheels, and drawing a plumb line on the photograph that originates at the axis of the contacting wheels, the line would pass through the C.G. Assuming that the "z" component lies directly above the "x and y" C.G. coordinates, the intersection of the drawn plumb line with the line which bisects the vehicle is the C.G. The "z" component then can be scaled from the photograph.

As a rule of thumb, most late-model passenger cars have a "z" coordinate about 2 feet from the ground. Regular sized pickup trucks, e.g. Ford 100, have a "z" coordinate located about 2.5 feet above the ground.

J. Moment of Inertia

The moment of inertia of a vehicle can be determined experimentally by suspending the vehicle from a single cable and variously spinning it and measuring the rate of rotation. This experimental method takes advantage of the relation:

$$F \times r = I \alpha \qquad \text{(xxviii)}$$

where

F = the applied force,
r = distance from the C.G.,
I = moment of inertia, and
α = angular acceleration.

In this case, the cross product between two vectors is used, which is denoted by "×".

In general, a constant force is applied at one of the far ends of the vehicle, perpendicular to the lever arm between the point of application and center of gravity of the vehicle. The final angular velocity, "wf", is measured and its value is divided by the time the force was applied to obtain "a," the angular acceleration. Thus, the moment of inertia is found by the following:

$$I = [F \times r] / [\omega f/t] \qquad \text{(xxix)}$$

In Section I of this chapter it was shown how to find the center of gravity by suspending the vehicle from a cable. Since the additional effort necessary to find "I" is rather small in comparison to the trouble of the set up, it is recommended that they both be done at the same time.

However, if a person is unable to undertake hoisting up a car at the end of a cable and spinning it, there are several methods that can be used to provide reasonable estimates of "I."

The first method assumes that the distribution of mass in the vehicle is more or less even over the whole projected x-y area and that the center of gravity is near the geometric center.

$$m_{areal} = m/A \qquad \text{(xxx)}$$

where

A = area of the vehicle projected onto the horizontal plane,
m = mass of the vehicle, and
m_{areal} = mass per unit area.

For a rectangular object the moment of inertia is:

$$I_{xx} = m_{areal}bh^3/12 = m_{areal}Ah^2/12 = mh^2/12 \qquad \text{(xxxi)}$$

where

b = width,
h = length, and
I_{xx} = moment of inertia about x-x axis.

Using the above method for a 1988 Buick Century, which is 189 inches long and 69 inches wide with a curb weight of 2,950 lbs, its moment of inertia would be calculated as follows:

$$m = 2{,}950 \text{ lbf}/(386 \text{ in/sec}^2) = 7.642 \text{ lb-sec}^2/\text{in}$$

$$I_{xx} = (7.642 \text{ lb-sec}^2/\text{in})(189 \text{ in})^2/12 = 22{,}750 \text{ lb-sec}^2\text{-in.}$$

A second method considers the loads carried by the axles as point masses located half the wheelbase from geometric center.

$$I_{xx} = (x)m(l/2)^2 + (1 - x)m(l/2)^2 = ml^2/4 \qquad \text{(xxxii)}$$

where

l = length of wheelbase.
x = fraction of mass carried by front axle, and
(1 - x) = fraction of mass carried by rear axle.

Using the same 1988 Buick Century as before, with a wheelbase of 105 inches, the "I" value is calculated as follows.

$$I_{xx} = (0.65)(7.642 \text{ lb-sec}^2/\text{in})(105 \text{ in}/2)^2 + (0.35)(\text{lb-sec}^2\text{-in})(105 \text{ in}/2)^2$$

$$I_{xx} = 21{,}063 \text{ lb-sec}^2\text{-in.}$$

A third method assumes that it is the engine that is primarily responsible for the uneven distribution of load between axles. Thus, the mass of the engine is deducted from the total vehicle mass and an

"I" value is computed for the "engineless" vehicle. The engine mass is then added in as a point mass.

$$I_{xx} = 2(1 - x)mh^2/12 + (2x - 1)m(l/2)^2 \qquad \text{(xxxiii)}$$

where

x = fraction of load on front axle,
h = overall length of vehicle,
l = wheelbase length, and
m = mass of vehicle.

Again, using the 1988 Buick Century, the following "I" value is computed.

$$I_{xx} = 2(0.35)(7.642 \text{ lb-sec}^2\text{-in})(189 \text{ in})^2/12$$

$$+ (0.30)(7.642 \text{ lb-sec}^2\text{-in})(105 \text{ in}/2)^2$$

$$I_{xx} = 15{,}923 \text{ lb-sec}^2\text{-in} + 6{,}319 \text{ lb-sec}^2\text{-in} = 22{,}242 \text{ lb-sec}^2\text{-in}.$$

When a vehicle is carrying an unusually heavy load in the trunk or elsewhere, the additional mass can be figured in the computation for "I" in the same way that the engine mass was done.

Of the three methods, the third seems to work best for passenger cars, where the mass distribution is fairly even except for the engine. The first method seems to work best for truck trailers and vehicles where the general mass distribution is relatively even and there are no large concentrations of mass. Method two works well when there are heavy loads over the axles, like in a loaded pickup truck (engine over front axle, load over rear). However, as demonstrated by the 1988 Buick Century calculations, all three methods give similar results.

One additional item to note is that the "I" values discussed above assumed that the x-x axis was in the geometric center of the vehicle. As such, it was assumed that rotation would be about the center of the vehicle. If, however, rotation occurs where the center of rotation is about the rear axle or the front axle, then the parallel axis theorem will have to be applied to properly adjust the "I" to the point of rotation. The parallel axis theorem is given below.

$$I_{ii} = I_{xx} + mr^2 \qquad \text{(xxxiv)}$$

where

I_{ii} = moment of inertia about axis i-i,

r = distance between axis i-i and axis x-x, and
m = mass of the vehicle.

K. Torque

Again, starting with the basic equation of engineering:

$$F = m\,a$$

Let us now apply the force not to an idealized ball or point of mass, but to a stiff object with some length. Further, let the force be applied at a point so that the line of action of the force does not pass through the center of mass. In that case, the following occurs.

$$T = F \times r = (ma) \times r = m\,(a \times r) \qquad \text{(xxxv)}$$

where
T = torque,
r = vector distance from center of mass,
F = force applied at "r",
m = total mass of item, and
a = acceleration in the direction of the force.

As before, the "×" denotes a cross product vector multiplication.

The rotation of a body around its center of mass can be constant or variable. When the rotation is constant, the rotation rate is usually given in radians per second. When the rotation rate varies, the rate of change is quantified by the angular acceleration, which is given in radians per second per second. The relationship between angular velocity, "w" and angular acceleration, "a" is given as follows.

$$\alpha = \Delta w / \Delta t \qquad \text{(xxxvi)}$$

where
Δ denotes an incremental change, and
t = time.

The tangential velocity at distance "r" from the center of mass is given by the following:

$$v_T = \omega \times r \qquad \text{(xxxvii)}$$

Similarly, the tangential acceleration at distance "r" from the center of mass is given by the following:

$$a_T = (d/dt)(\omega \times r) = \alpha \times r \qquad \text{(xxxviii)}$$

Substituting that back into equation (xxxv) gives the following:

$$T = F \times r = m(\alpha \times r) \times r \qquad \text{(xxxix)}$$

Collecting terms and simplifying gives:

$$T = F \times r = (mr^2)\,\alpha = I_{xx}\,\alpha \qquad \text{(xl)}$$

where I_{xx} = the moment of inertia of a point mass about the axis of rotation.

If the mass is distributed over an area instead of at a point, the moment of inertia about the center of mass is used, provided the rotation is about the center of mass. If the rotation is about some other axis, then the parallel axis theorem is applied (equation (xl)) to determine the moment of inertia about that particular axis.

L. Angular Momentum Equations

Like its linear acceleration counterpart, rotational acceleration is the time derivative of rotational velocity. Like its linear counterpart force, torque is also equal to a time derivative of velocity.

$$T = I_{xx}\,(d/dt)(\omega) = (mr^2)(d/dt)(\omega) = m\,[(d/dt)(\omega)] \times r \times r \qquad \text{(xli)}$$

In the above, it is assumed that the "I_{xx}" is a point mass at a distance "r" from the axis of rotation.

Since $(d/dt)(\omega) \times r = (d/dt)(v_T)$, then

$$T\,(dt) = I_{xx}\,(d\omega) = m(dv_T) \times r \qquad \text{(xlii)}$$

where v_T = tangential velocity at point "r" on the body.

The indefinite integration of both sides of equation (xlii), as shown below, yields a term called the rotational impulse.

$$R = \int[I_{xx}(d\omega)] = \int[m(dv_T) \times r] \qquad \text{(xliii)}$$

$$R = I_{xx}[\omega_2 - \omega_1] = [m(v_{T2} - v_{T1}) \times r]$$

where R = the rotational impulse.

The term "$I\omega$" is referred to as the angular momentum.

Equation (xliii) is enormously useful in dealing with vehicular collisions which produce rotation. The left side of the equation indicates that the rotational impulse is simply the change in angular momentum. The right side of the equation indicates that the rotational pulse is equal to the change in linear momentum at point "r" about the rotational axis.

To demonstrate the usefulness of equation (xliii), consider the following collision. Car "A" is traveling due south. Car "B" is traveling due east, but is stopped at an intersection waiting for the light to change. Car "B" creeps forward slowly into the intersection while it is waiting for the light to change. Car "A" drives through the intersection and impacts car "B" at its front left corner, causing it to spin or yaw about the rear wheels. Car "A" does not spin, but is stopped by the collision at the point of impact.

The forward linear momentum of car "A" has been transferred to car "B" causing it to spin. The momentum of car "A" was "$m_A v_A$." This momentum was then transferred to car "B" at a point "r" from the axis of the rear wheels. The relationship of the spin, or angular momentum, of car "B" to the loss of the forward linear momentum of car "A" is given by:

$$(I_{Bxx} + m_B r^2)\omega = m_A v_A \times r$$

where

r = distance from the rear axle to the front left corner of car "B"
I_{Bxx} = moment of inertia about the center of mass of car "B"
$m_A v_A$ = forward moment of inertia of car "A"
ω = spin or yaw of car "B" just after impact is completed.

It should be noted that the speed "v_A" of car "A" is tangential to the y-y or long axis of car "B". If the impact had been at an angle other than 90°, it would be necessary to take the normal component of the velocity vector with respect to car "B's" y-y axis. Alternatively, this is expressed as:

$$v_A \times [r/(|r|)] = |v_A| \sin \phi = v_{AT}$$

where

ϕ = the acute angle between the axis y-y of car "B" and the velocity vector of car "A", and

$[r/(|r|)]$ = the unit directional vector "r" .

M. Solution of Velocities Using the Coefficient of Restitution

As already noted, equation (xvii) defines the coefficient of restitution as:

$$e = I_{res} / I_{def} = [v_{B2} - v_{A2}] / [v_{A1} - v_{B1}]$$

If the above expression for "e" is alternately solved for "v_{A1}" and "v_{B1}", and the results are then substituted back into the conservation of momentum equation, equation (vi), then the following equations are obtained:

$$v_{A1} = [(em_A - m_B)/e(m_A + m_B)]v_{A2} + [(em_B + m_B)/e(m_A + m_B)]v_{B2} \qquad \text{(xliv)}$$

$$v_{B1} = [(em_A + m_A)/e(m_A + m_B)]v_{A2} + [(em_B - m_A)/e(m_A + m_B)]v_{B2} \qquad \text{(xlv)}$$

Equations (xliv) and (xlv) indicate that if the velocities just after collision are known, and the "e" value of the collision is known, then the individual velocities of each body prior to impact can be calculated. Inspection of these equations find that both expressions are linear vector equations; both "v_{A1}" and "v_{B1}" are resultants of the addition of two vector quantities. This solution form lends itself well to graphic solution.

An interesting point is worth noting about equations (xliv) and (xlv). If the two equations are written in the following form:

$$v_{A1} = [a]v_{A2} + [b]v_{B2} \qquad \text{(xlvi)}$$

$$v_{B1} = [c]v_{A2} + [d]v_{B2}$$

where a, b, c, and d are scalars, then the following relation holds between the scalars:

$$a + b = c + d = 1 \qquad \text{(xlvii)}$$

The above relationship can be very useful in checking the computations of the scalars, and can be used as a calculational shortcut. Using this relationship, only one scalar in each equation needs to be computed, and the other scalar can be quickly determined by inspection. The relationship can also save programming lines if the equations are programmed for solution on a computer.

If "e = 1", a fully elastic collision, it is seen that equations (xliv) and (xlv) reduce to the following:

$$v_{A1} = [(m_A - m_B)/(m_A + m_B)]v_{A2} + [(2m_B)/(m_A + m_B)]v_{B2} \quad \text{(xlviii)}$$

$$v_{B1} = [(2m_A)/(m_A + m_B)]v_{A2} + [(m_B - m_A)/(m_A + m_B)]v_{B2} \quad \text{(xlix)}$$

In this special case where the collision is wholly elastic, the scalar coefficients in equation (xlvi) not only satisfy a + b = c + d = 1, but also the following:

$$a = -d \quad \text{(l)}$$

$$b + c = 2$$

$$c - a = 1$$

$$b - d = 1$$

If "e = 0", a fully plastic collision, it is seen that equations (xliv) and (xlv) are undefined due to the division by zero. However, when "e = 0", it is automatically known that "$v_{A2} = v_{B2} = u$."

In a similar fashion, if equation (xvii), the relation for the coefficient of restitution, is alternately solved for "v_{A2}" and "v_{B2}", and the results are substituted back into the equation for the conservation of momentum, equation (vii), then the following is obtained:

$$v_{A2} = [(m_A - em_B)/(m_A + m_B)]v_{A1} + [(m_B + em_B)/(m_A + m_B)]v_{B1} \quad \text{(li)}$$

$$v_{B2} = [(m_A + em_A)/(m_A + m_B)]v_{A1} + [(m_B - em_A)/(m_A + m_B)]v_{B1} \quad \text{(lii)}$$

Equations (li) and (lii) indicate that if the velocities just prior to collision are known, and if the "e" value for the collision is known, then the individual velocities after impact can be calculated. This is also a very useful analytical tool. The velocities of the two vehicles prior to impact, as reported by the drivers or witnesses, can be substituted into "v_{A1}" and "V_{B1}". Then, the post-impact velocities can be solved. The calculated post-impact velocities can then be compared to the actual accident scene information. If they match reasonably well, the drivers' accounts are validated. If they do not match, then one or both of the claimed pre-impact velocities are incorrect.

As before, these equations can also be arranged in the form:

$$v_{A2} = [w]v_{A1} + [x]v_{B1} \tag{liii}$$

$$v_{B2} = [y]v_{A1} + [z]v_{B1}$$

As before, it is found that the following relationship holds for the scalars:

$$w + x = y + z = 1 \tag{liv}$$

In a fully elastic collision, equations (li) and (lii) simplify to the following:

$$v_{A2} = [(m_A - m_B)/(m_A + m_B)]v_{A1} + [(2m_B)/(m_A + m_B)]v_{B1} \tag{lv}$$

$$v_{B2} = [(2m_A)/(m_A + m_B)]v_{A1} + [(m_B - m_A)/(m_A + m_B)]v_{B1} \tag{lvi}$$

In a fully elastic collision, it is seen that in addition to equation (liv), the additional relations hold between the scalars:

$$w = -z \tag{lvii}$$

$$x + y = 2$$

$$y - w = 1$$

$$x - z = 1$$

The reader will undoubtedly note the similarity of the relations in equation (lvii) with those of equation (l).

Similarly, when the collision is fully plastic and "e = 0", then the following holds:

$$v_{A2} = v_{B2} = [(m_A)/(m_A + m_B)]v_{A1} + [(m_B)/(m_A + m_B)]v_{B1} \tag{lviii}$$

N. Estimation of a Collision Coefficient of Restitution from Fixed Barrier Data

When a vehicle impacts a fixed barrier, it is assumed that the fixed barrier does not crush or otherwise absorb any significant amount of kinetic energy from the vehicle. The fixed barrier pushes back against the vehicle during impact with a force equal to that being applied by the vehicle. In essence, since no work can be done on the fixed barrier, the vehicle does work on itself.

Earlier, equation (xxv) was derived to show the relationship between energy dissipated by the collision and the coefficient of restitution. For convenience, the equation is shown again below.

$$E_{Loss} = [(1 - e_{res}^2)/2][(m_A m_B)/(m_A + m_B)][v_{A1} - v_{B1}]^2 \qquad \text{(xxv)}$$

If it is assumed that a fixed barrier has an essentially "infinite" or immovable mass, and the wall has no initial speed prior to the collision, then the equation reduces to the following:

$$E_{Loss} = [(1 - e_{res}^2)/2][m_A][v_{A1}^2] \qquad \text{(lix)}$$

When "e_{res} = 0", a fully plastic collision, all of the initial kinetic energy is dissipated as work in crushing the vehicle. When "e_{res} = 1.0", none of the initial kinetic energy is dissipated as work; all of it goes into the rebound of the vehicle.

While not exactly contributing to the point at hand, the following aside may be interesting to the reader. By substituting equation (xviii) into (lix), a function for collision energy loss against a fixed barrier collision can be derived that is solely dependent upon initial speed:

$$E_{Loss} = [(1 - e^{-0.130s})/2][m_A][s_{A1}^2] \qquad \text{(lx)}$$

where

s = velocity of vehicle "A" given in mph, and
e = 2.718 etc.

An alternative version of the above equation where velocity is given in ft/sec is as follows:

$$E_{Loss} = [(1 - e^{-0.089v})/2][m_A][v_{A1}^2] \qquad \text{(lxi)}$$

However, getting back to the point of this section, if it is assumed that crush caused by collision with the fixed barrier is proportional to the change in kinetic energy of the vehicle, then:

$$s_A \cdot F = \tfrac{1}{2}\, m_A(\Delta v_A{}^2) = U = E_{Loss} \tag{lxii}$$

where
- U = work done in crushing the vehicle,
- s_A = amount of crush in the vehicle,
- F = the average force applied to the vehicle during collision with the fixed barrier, and

$\Delta v_A{}^2 = v_{A2}{}^2 - v_{A1}{}^2$.

When two vehicles collide head on, the total work done in crushing is as follows:

$$U = \tfrac{1}{2}\, m_A(\Delta v_A{}^2) + \tfrac{1}{2}\, m_B(\Delta v_B{}^2) = s_A \cdot F + s_B \cdot F \tag{lxiii}$$

The force "F" applied between the two vehicles during collision is the same in magnitude but opposite in sign.

Now, if in the above equation, vehicle "B" is sitting still with respect to vehicle "A" (i.e., $\Delta v_B{}^2 = 0$), then the crush damage to both vehicles would be due to the action of vehicle "A" alone. As can be seen in comparing equations (lxii) and (lxiii), this means that the crush on vehicle "A" in colliding with vehicle "B" would be about half that of vehicle "A" colliding with a fixed barrier at the same speed.

$$\tfrac{1}{2} = v_{Aeq}{}^2 \,/\, v_{A1}{}^2 \text{ or } v_{Aeq} = (0.707)v_{A1} \tag{lxiv}$$

where
- v_{Aeq} = impact velocity of "A" with a fixed barrier, and
- v_{A1} = impact velocity of "A" with vehicle "B" which is sitting still.

Thus, the coefficient of restitution for the above collision situation can be estimated by converting "$(0.707)v_{A1}$" into mph units, and substituting into equation (xviii).

In a similar fashion, equivalent fixed barrier impact velocities can be derived for other situations. For example, if the two vehicles have about the same mass and collide with the same velocity head on, then the equivalent fixed barrier impact velocity is equal to the actual impact velocity.

O. Discussion of Coefficient of Restitution Methods

In general, the pre-impact velocities in a two-vehicle collision can be solved by proceeding as follows:

1. A reasonable "guess" as to the velocities of the vehicles is made. It is important to note that the "guess" is for the central impact velocities. If the collision was oblique, then it may be necessary to separate out the central velocity vector components, and handle the others separately.
2. A tentative value for "e" is calculated from the first "guess."
3. Using the tentative value for "e", equations (xliv) and (xlv) are solved.
4. A new value of "e" is calculated based upon the outcome of equations (xliv) and (xlv).
5. If the new value of "e" matches the tentative value of "e" first assumed, and the calculated velocities reasonably match the "guessed" values, then a reasonable solution has been obtained.
6. If the variance is too great, then a new "guess" of the velocities is made and the process is repeated until a reasonable convergence occurs.

However, in solving pre-impact velocities using the above method, equations (xliv) and (xlv) become extremely sensitive to small variations in the value of "e" when "e" is very small. Of course, when "e" is small, the collision is mostly plastic, and has occurred at a relatively high speed.

Thus, when "e" is very small and subject to imprecision, an alternative approach may be advisable.

1. Using equations (li) and (lii), the pre-impact velocities are assumed, or taken to be the values as stated by the drivers or witnesses.
2. An "e" value is then determined from equivalent barrier impact velocity computations, or simply assumed.
3. The post-impact velocities are then calculated.
4. From the actual police report data, the post-impact velocities are determined. These velocity vectors are then compared to

the calculated velocity vectors. If they reasonably match, a satisfactory solution has been found.

5. If the post-impact velocities do not reasonably match, new pre-impact velocity vectors are assumed and the process is repeated until convergence is achieved.

Chapter 9: Example Problems

Problem 1

A rubber ball being dropped straight down onto a floor and rebounding is one of the more common examples of an elastic collision. The ball falls from a certain height and attains maximum velocity just prior to impact. The ball then rebounds upwards with full dimensional restoration. The maximum velocity after impact occurs just after it leaves the floor. A rubber ball can do this many times without any apparent damage or permanent distortion of dimensions.

In this case, a rubber ball filled with air, similar to that used to play racquetball, was dropped from three different heights onto a firm hardwood floor. The ball was 2.25 inches in diameter, and weighed 45 grams. The height of rebound was then measured using a carpenter's tape measure. Table 1, below, lists the three drop heights and the average rebound height corresponding to each elevation. Analyze this in terms of an elastic impact.

Table 1. Drop and Rebound Heights of Rubber Ball

Drop Height, in.	Rebound Height, in.
36	24
72	46
96	63

Solution

The maximum velocity at impact when the ball is dropped straight down, neglecting air friction, is given by:

$$v = -(2gh)^{0.5}$$

where

g = gravitational acceleration, and
h = height of drop.

Similarly, the velocity just after rebound can be calculated with the same formula given above if the rebound height is substituted for "h" and the sign is changed to account for the upward direction.

In this simple case where there is only velocity in the vertical direction, and the second body (the floor) remains motionless, equation (xvii) for the coefficient of restitution reduces to the following:

$$e = (-v_2)/(v_1)$$

Substituting the previous relationships between velocity and height into the above gives the following:

$$e = (h_2/h_1)^{0.5} \quad \text{or} \quad e^2 = h_2/h_1$$

where
h_1 = initial drop height and
h_2 = height of rebound.

In the three drops listed in Table 1, the coefficients of restitution are calculated to be as shown in Table 2, below.

Table 2. Coefficients of Restitution of Rubber Ball

Drop Height, in.	Coefficient of Restitution
36	0.82
72	0.80
96	0.81

The above table shows that, within experimental error, the coefficient of restitution remained constant, about 0.81, over the range tested. While it might be expected that a rubber ball would exhibit more or less constant elastic properties over a wide range of drop heights, the relatively low coefficient of restitution value of 0.81 may be surprising as this is considered to be an elastic impact. Perhaps a value closer to 1.0 would have been initially anticipated.

The amount of energy lost in the impact with the hardwood floor can be calculated from equation (xxix) as follows:

$$E_{Loss} = \tfrac{1}{2}(m)(v_1^2)(1 - e^2)$$

Evaluation of the last factor in the above equation finds that the energy loss in a single impact is about 34% of the total kinetic energy just before impact. If the rubber ball were allowed to bounce "n" times, then the amount of energy lost after "n" bounces would be:

$$E_{Loss} = (mgh_{drop})(1 - e^{2n})$$

where

h_{drop} = height of first drop.

The above is true because:

$ev_1 = v_2$ by definition,

$ev_2 = v_3 = e^2v_1$ after two bounces,

$ev_3 = v_4 = e^2v_2 = e^3v_1$ after three bounces, and

$\ldots\ldots\; v_{n+1} = e^nv_1$ after "n" bounces.

Since no noticeable permanent damage occurs to the rubber ball, it is assumed that the kinetic energy is dissipated by internal frictional heating of the rubber ball, a little bit with each bounce. Since the specific heat of rubber is about 0.48, then the temperature increase of the ball would be:

$$E_{Loss} = C_pm[\Delta T]$$

$$\Delta T = E_{Loss}/(C_pm)$$

where

C_p = heat capacity of rubber at constant pressure,
m = mass of ball, and
T = temperature.

It is noted that the increase in temperature with each bounce can be calculated by substituting the equation for energy loss for "n" bounces in the above.

In the ball used in the experiment, the amount of temperature increase that would occur when the ball is dropped 72 inches and is allowed to bounce until it stops is as follows:

$\Delta T = [(0.099 \text{ lbf})(6 \text{ ft})]/[(0.48 \text{ BTU/lbm°R}) (0.099 \text{ lbm})][778 \text{lbf-ft/BTU}]$

$\Delta T = 0.01°R = 0.01°F$

Thus, the rubber ball would only become 0.01 degrees Fahrenheit warmer if it were dropped from a height of 72 inches and bounced until it came to a stop. A single bounce from 72 inches would have caused only a 0.0034 degree increase.

The following can be generalized from the above solution:

1. The coefficient of restitution between two items holds more or less constant within the elastic impact range.

2. The temperature increase in the materials caused by elastic internal friction is trivial.

Equation (xviii) developed previously in the chapter related the coefficient of restitution of a vehicle impacting an immovable wall to its forward speed. If "s" is set equal to 2.5 mph, the current minimum impact speed at which a bumper is supposed to remain undamaged, the corresponding "e" value is calculated to be 0.85. This is only 5% more than the value of "e" experimentally determined for the rubber ball.

Problem 2

In an actual crash test, a 1979 Ford Mustang was driven at 35 mph into an immovable brick wall. The car recoiled 18.3 inches from the wall after impact. Determine its pre-impact and post-impact speeds.

Solution

When a vehicle collides with a firm object and suddenly changes from forward to backward motion, the drive wheels are not able to rotate freely. They will often be held immobile by the drive train gearing, which is set for forward motion. Thus, the drive wheels are locked.

A 1979 Mustang has rear-wheel drive, and the rear of the vehicle carries about 40% of the curb weight. Thus, the work done during recoil by the locked rear wheels is given by:

$U = (0.40) \, W \, f \, d$

where

U = work done by locked rear wheels moving backwards,
W = curb weight of vehicle,
f = coefficient of friction, 0.70, and
d = distance moved, 18.3 inches.

The velocity just after impact is then given by:

$$v_2 = [2(0.40)\, g\, f\, d]^{0.5}$$

$$v_2 = [2(0.40)(32.17\ \text{ft/sec}^2)(0.7)(1.525\ \text{ft})]^{0.5} = 5.24\ \text{ft/sec}$$

$$v_2 = 3.57\ \text{mph}$$

Given that the initial velocity was 35 mph before impact, or 51.33 ft/sec, then the coefficient of restitution is as follows:

$$e = -v_2/v_1 = 5.24/51.33 = 0.102$$

Alternately applying equation (xviii) to estimate "e" from the initial speed, it is found that:

$$e_{res} = e^{-0.065(35)} = 0.103$$

Using the above value of "e" to determine the post-impact speed, it is found that "v_2" would be 5.29 ft/sec or 3.60 mph.

By evaluating the factor "$(1 - e^2)$", it is found that 98.96% of the initial kinetic energy of the Mustang was dissipated in the impact. For all practical purposes, this was a non-elastic impact.

Problem 3

In actual crash tests•, a 1980 Citation was crashed into an immovable wall at 40 mph and rebounded 7 inches. A 1975 Honda Civic was similarly crashed at a speed of 19.7 mph and rebounded 2.8 inches, and a 1986 Subaru station wagon was crashed at a speed of 30 mph and rebounded 4.0 inches. Compare "e" values from basic calculations to those generated by equation (xviii).

(•Note: crash data from "*Elastic Properties of Selected Vehicles*" by Navin, MacNabb and Miyasaki; Accident Research Team, University of British Columbia, SAE paper 880223, table 1.)

Solution

The 1980 Citation also has rear wheel drive with 40% of the vehicle's weight on the rear wheels. Thus with a rebound of 7 inches, and assuming a locked rear wheel condition, the velocity after impact would be 3.24 ft/sec or 2.21 mph. The coefficient of restitution is then 0.055, and 99.70% of the kinetic energy would have been consumed in the impact.

Applying equation (xviii) and using "s = 40 mph" for the above, the calculated coefficient of restitution is then 0.074, and 99.45% of the kinetic energy would have been consumed in the impact. A rebound speed of 2.97 mph would then be expected along with a rebound of 8.4 inches.

For the 1975 Honda Civic, the same assumptions are used except that the Honda has a front-wheel drive. Thus, a factor of 0.65 is used instead of 0.4 to account for front-wheel lock up. The velocity after impact is calculated to be 2.61 ft/sec or 1.78 mph. The coefficient of restitution is then calculated to be 0.090, and 99.19% of the kinetic energy was consumed in the crash.

Applying equation (xviii) and using "s = 19.7 mph", the calculated coefficient of restitution is then 0.278. Using this, a rebound velocity of 5.47 mph would be predicted, with an associated rebound of 2.2 ft. About 92.27% of the kinetic energy would be consumed in the impact.

Finally, the Subaru station wagon is also a front-wheel drive vehicle, thus a factor of 0.65 is used. A rebound speed of 3.12 ft/sec or 2.13 mph is calculated and the coefficient of restitution is 0.071. About 99.50% of the kinetic energy would have been consumed in the impact.

Applying equation (xviii) and using "s= 30 mph", the coefficient of restitution is determined to be 0.142. Using this, a rebound speed of 4.27 mph or 6.26 ft/sec is calculated, with a rebound of 1.34 ft. About 97.98% of the kinetic energy would have been consumed in the impact.

As is demonstrated in these sets of comparative figures, the use of equation (xviii) often gives results that apparently vary with actual data. However, this variance is actually quite small when its basis is understood.

The factor "$(1 - e^2)$" used in various equations indicates the fraction of energy dissipated in the impact. If "e" is set equal to 0.10, then $(1 - e^2) = 0.99$. If "e" is equal to 0.141, then $(1 - e^2) = 0.98$. In other words, in this regime a decrease in energy dissipation by impact of only 1% will cause a 41% increase in the value of "e," and a corresponding 41% increase in the rebound velocity. And, of course, a 41% increase in rebound velocity will cause a near doubling of the rebound distance.

Thus, very small changes in the amount of crash energy dissipated can cause large changes in rebound parameters in this regime.

Taken in this context, it can be seen that equation (xviii) works very well. For example, with the 1980 Citation, the apparent rebound "errors" between actual data, and predictions based on equation (xviii) amounted to 0.25% of the total dissipated impact energy. With the 1986 Subaru station wagon, the "errors" amounted to 1.52%, and with the 1975 Honda Civic the "errors" come to 6.92%. The larger discrepancy in the 1975 Honda Civic indicates that the Honda may be "softer" than most cars and is more easily crushed by impact.

Problem 4

A van was traveling due north (0 degrees) through an intersection. A motorcycle and rider was traveling due west through the same intersection and collided with the van, at its right front wheel. The van's trajectory then deviated to a direction 30 degrees west of north after impact, and skidded to a stop in 29 feet. The motorcycle and rider basically bounced off of the van and landed close to the point of impact.

The driver of the van indicated he did not see the motorcycle coming, and that he applied the brakes directly after impact without making any steering maneuvers. A witness at the scene indicated that the motorcycle may have been traveling 40 mph or more.

The van's curb weight was 3,100 lbs. With two adult passengers, the total weight of the van was estimated to be 3,450 lbs. The curb weight of the motorcycle was 300 lbs; with the rider, its total weight was 475 lbs. The speed limit for the area was 25 mph, and the road surface was dry asphalt.

From the above, determine the speeds of the van and motorcycle prior to impact.

Solution

Since the van skidded 29 ft after impact, its speed just after impact is determined as follows:

$$v_2 = [2g(29 \text{ ft})(0.7)]^{0.5} = 36 \text{ ft/sec} = 24.6 \text{ mph}$$

The above speed is directed along the trajectory of the van after impact. In cartesian vector notation, with the "j" axis being due north, the velocity vector of the van after impact is as follows:

$$v_2 = -(36 \text{ ft/sec})(\sin 30)i + (36 \text{ ft/sec})(\cos 30)j$$

$v_2 = -(18\ \text{ft/sec})i + (31.2\ \text{ft/sec})j$

The momentum vector for the van after impact then is the post-impact velocity vector times the total mass of the van.

$m_{van} = (3{,}450\ \text{lb})/(32.17\ \text{ft/sec}^2) = 107.2\ \text{lb-sec}^2/\text{ft}$

$M_{van,2} = -(1{,}930\ \text{lb-sec})i + (3{,}344\ \text{lb-sec})j$

Now, the motorcycle and rider had essentially no post-impact velocity vector, and had therefore no post-impact momentum vector. Prior to impact, the motorcycle was traveling in the -i direction only, and the van was traveling in the +j direction only. When the motorcycle collided with the van, all of the momentum of the motorcycle and rider was transferred to the front of the van. This caused the front of the van to shift to the left (west). With the front of the van moving in a new direction, the rear of the van simply followed. Thus, some of the -i directed momentum of the van after impact is due to momentum transfer from the motorcycle and rider, and some is due to side forces on the tires in following the front of the van in its new direction.

Since the front end of the van carried 60% of the weight of the vehicle, then 60% of the -i directed momentum of the van after impact is due to momentum transfer from the motorcycle and rider.

$m_{motorcycle} = (475\ \text{lb})/(32.17\ \text{ft/sec}^2) = 14.77\ \text{lb-sec}^2/\text{ft}$

$M_{van,1} + M_{motorcycle,1} = M'_{van,2}$

where

$M'_{van,2}$ = momentum of van after impact with the 60% multiplier factored into the -i direction.

$-(14.77\ \text{lb-sec}^2/\text{ft})(V_1)i + (107.2\ \text{lb-sec}^2/\text{ft})(v_1) = M'_{van,2}$

$M'_{van,2} = -(0.60)(1{,}930\ \text{lb-sec})i + (3{,}344\ \text{lb-sec})j$

Thus,

$v_1 = (3{,}344\ \text{lb-sec})/(107.2\ \text{lb-sec}^2/\text{ft}) = 31.2\ \text{ft/sec} = 21.2\ \text{mph}$

which is the nominal northbound speed of the van prior to impact, and

$V_1 = (0.60)(1{,}930\ \text{lb-sec})/(14.77\ \text{lb-sec}^2/\text{ft}) = 78.4\ \text{ft/sec} = 53.4\ \text{mph}$

which is the westbound speed of the motorcycle and rider before impact.

However, there is one last correction to be made to the speed of the van just prior to impact. When 60% of the -i directed momentum was attributed to transfer from the motorcycle and rider, then the remaining 40% must have been due to the original momentum of the van prior to impact. Thus, the momentum of the van after impact due to its speed before impact is as follows:

$$M''_{van,2} = -(0.40)(1{,}930 \text{ lb-sec})\mathbf{i} + (3{,}344 \text{ lb-sec})\mathbf{j}$$

$$M''_{van,2} = 3{,}431 \text{ lb-sec in a direction } 13^\circ \text{ west of north.}$$

Thus, the speed of the van prior to impact with the motorcycle and rider was as follows:

$$v'_1 = (3{,}431 \text{ lb-sec})/(107.2 \text{ lb-sec}^2/\text{ft}) = 32.0 \text{ ft/sec} = 21.82 \text{ mph}$$

Thus, just prior to impact, the van was traveling northward at 21.82 mph.

The above analysis indicates that the van was traveling at a proper speed, the motorcycle was speeding excessively, and the witness was reasonably correct in his observations.

There is one final note to consider. If just the speed of the van after impact had been computed from the skid marks and the momentum analysis had not been done, it might have been concluded that the van was traveling faster than otherwise determined. The -i directed speed component added by the impact of the motorcycle made the post-impact speed of the van a little faster.

Problem 5

A pickup truck was traveling due south through an intersection. A northbound passenger car approached the same intersection from the south and made a left turn to proceed west. As the passenger car made the left turn, it collided with the pickup truck.

After impact, the pickup truck veered so that instead of going due south, it was traveling in a direction 10 degrees west of south and skidded 38 feet to a stop. After impact, the passenger car skidded 4 feet to a stop in a direction 60 degrees north of west. Examination of the damages found that "slide by" contact, where the pickup would have "wiped" across the front of the passenger car, was minimal.

The weight of the pickup truck was 3,100 lbs and the weight of the two passengers was 400 lbs. The weight of the passenger car was 2,800 lbs, and the weight of the driver was 130 lbs. The roadway surface was dry asphalt. The speed limit for the area was 35 mph.

What were the pre-impact speeds of the vehicles, and what was the approximate angle of impact of the passenger car with respect to the pickup?

Solution

If the pickup truck skidded 38 feet to a stop, the speed after impact can be determined by the ever faithful skid equation.

$$|v_{A2}| = [2g(0.70)(38\ \text{ft})]^{0.5} = 41.36\ \text{ft/sec} = 28.2\ \text{mph}$$

Applying cartesian vector notation, and using due north as +j and due east as +i, then the velocity vector for the pickup truck after impact was as follows:

$$v_{A2} = -(41.36\ \text{ft/sec})(\sin 10°)i - (41.36\ \text{ft/sec})(\cos 10°)j$$
$$v_{A2} = (-7.18\ \text{ft/sec})i + (-40.73\ \text{ft/sec})j$$

Similarly, the velocity of the passenger car after impact was as follows:

$$|v_{B2}| = [2g(0.7)(4\ \text{ft})]^{0.5} = 13.42\ \text{ft/sec}.$$

The vector representation of the post-impact speed of the passenger car is as follows:

$$v_{B2} = (-13.42\ \text{ft/sec})(\sin 60°)i + (13.42\ \text{ft/sec})(\cos 60°)j$$
$$v_{B2} = (-6.71\ \text{ft/sec})i + (11.62\ \text{ft/sec})j$$

The respective masses of the two vehicles are:

$$m_A = (3{,}500\ \text{lbs})/(32.17\ \text{ft/sec}^2) = 108.8\ \text{lb-sec}^2/\text{ft}$$
$$m_B = (2{,}930\ \text{lbs})/(32.17\ \text{ft/sec}^2) = 91.1\ \text{lb-sec}^2/\text{ft}$$

If it is assumed that the southerly velocity of the pickup truck was not significantly affected by the impact, then $v_{A1} = -(40.73\ \text{ft/sec})j$. This assumption can be verified by examination of the vehicles to determine if there was significant "slide by" frictional contact. In this case, it was not significant. Thus, the following momentum balance can be made:

$[108.8 \text{ lb-sec}^2/\text{ft}][((-7.18 \text{ ft/sec})i + (-40.73 \text{ ft/sec})j] + [91.1 \text{ lb-sec}^2/\text{ft}]$

$[(-6.71 \text{ ft/sec})i + (11.62 \text{ ft/sec})j] = [108.8 \text{ lb-sec}^2/\text{ft}][(-40.73 \text{ ft/sec})j] +$

$[91.1 \text{ lb-sec}^2/\text{ft}][v_{B1}].$

$$v_{B1} = -(15.29 \text{ ft/sec})i + (11.62 \text{ ft/sec})j$$

$$|v_{B1}| = 19.2 \text{ ft/sec at an angle of } 37.2° \text{ north of west.}$$

Since the passenger car was in the process of making a left turn, the calculated impact speed of 19.2 ft/sec or 13 mph seems reasonable. It is typical of the speed of a vehicle executing a conservative 90° left turn. From the angle of contact, it appears that the passenger car was in the last part of its turn when impact occurred.

Problem 5

Two vehicles collided head on at the crest of a hill. Neither driver could see the other approach due to the slope of the hill, and both vehicles were driving in the center of the road. The driver of vehicle "A" reported he was traveling eastbound at 35 mph. The driver of vehicle "B" reported she was traveling westbound at 40 mph. Vehicle "A" was a compact car weighing 2,200 lbs. Vehicle "B" was a small pickup truck that weighed 3,200 lbs.

Determine, using coefficient of restitution analysis, the expected post-impact reactions of the two vehicles. From this, determine how a person could tell, by an examination of the accident scene, if either of the drivers reported their velocity inaccurately.

Solution

The first thing to do is to apply a simple momentum equation to the collision. It is initially assumed that the collision is basically plastic, and that the two vehicles moved more or less together after impact.

Since the collision is along one axis, the eastward direction is considered positive.

$$m_A v_{A1} - m_B v_{B1} = (m_A + m_B)u$$

$$u = -9.44 \text{ mph} = -13.9 \text{ ft/sec.}$$

The above result indicates that the two vehicles would move westward after impact at 9.44 mph. If both vehicles were braking, the skid distance would be about 4.26 ft, assuming a coefficient of friction of 0.7.

The second part of the analysis is a little more involved. First, the total kinetic energy of both vehicles before impact is summed as follows:

$$KE_1 = ½ m_A(v_{A1}^2) + ½ m_B(v_{B1}^2) = 261{,}282 \text{ lb-ft.}$$

One way to consider an equivalent collision of two vehicles is to imagine that both vehicles are driving side by side at the same speed and then simultaneously impacting a brick wall, i.e., a fixed barrier. The impact velocity at which the same amount of kinetic energy would be dissipated by crush in both vehicles is the equivalent fixed barrier velocity for the collision of these two vehicles.

$$v_{eq} = [(KE_1)/(m_A + m_B)]^{1/2} = 55.8 \text{ ft/sec} = 38 \text{ mph}$$

Substituting this into equation (xviii) gives an "e" value of 0.084. Applying the definition for the coefficient of restitution, it is found that:

$$[v_{B2} - v_{A2}] = 9.28 \text{ ft/sec.}$$

This result indicates that the two vehicles do not "stick" together after impact, but have a small relative velocity between them.

The basic momentum equation solved in the first part describes how the center of gravity between both vehicles moves after impact. The results of part two describe how the two vehicles separate relative to each other. By combining the two, an exact solution for the motion of both vehicles just after impact can be obtained.

$$(m_A + m_B)u = m_A v_{A2} + m_{B2}$$

where

$$[v_{B2} - v_{A2}] = 9.28 \text{ ft/sec.}$$

$$(167.9 \text{ lbf-sec}^2/\text{ft})(-13.9 \text{ ft/sec}) =$$

$$(68.4 \text{ lbf-sec}^2/\text{ft})(v_{A2}) + (99.5 \text{ lbf-sec}^2/\text{ft})(9.28 \text{ ft/sec} + v_{A2})$$

$$v_{A2} = -19.4 \text{ ft/sec} \qquad v_{B2} = -10.1 \text{ ft/sec}$$

Thus, at a coefficient of friction of 0.7, vehicle "A" would have skidded 8.4 ft westward, and vehicle "B" would have skidded 2.3 ft westward. The resulting separation between the two vehicles would have been 6.1 ft.

The main points in determining whether or not the drivers' reported speeds are accurate are:

1. The position of the point of impact with respect to the final position of the vehicles. The center of gravity of the combined two vehicles in their final positions should be a few feet westward of the point of impact.
2. The amount of rebound of the compact. The compact should have rebounded more than the pickup, but both vehicles should be west of the point of impact.
3. Separation between the two vehicles should be about 6 feet.

If any of the above points are violated by inspection of the scene, then the speeds reported by the drivers may be incorrect.

For example, if the actual speeds were 50 mph for the compact and 40 mph for the pickup (as reported by the driver), then the center of gravity between the two vehicles as found after impact would be almost over the point of impact. This is because the after-impact speed (assuming "e" = 0) would be 3.3 mph westward. At such a speed, and with a coefficient of friction of 0.7, the skid distance would be only 0.5 ft. Assuming an "e" value of 0.054 (an equivalent impact speed of 45 mph), then the speed of separation between the two vehicles after impact would be 7.1 ft/sec. Assuming that the pickup would not significantly move from the point of impact, the separation between the two vehicles would be only 1.1 ft.

CHAPTER 10

Energy Methods

A. General

In previous chapters, some aspects of energy methods have already been discussed. In Chapter 2, it was shown how the initial kinetic energy of a vehicle can be dissipated by irreversible work done during braking. In both Chapters 2 and 3, the interplay between kinetic energy and potential energy was examined in falls and elevation changes. In Chapter 9, it was shown how the irreversible work done in crushing a vehicle during impact relates to the degree of elasticity or plasticity of the collision. In all of these applications, the fundamental assumption is the conservation of energy.

This chapter discusses in a more general way how the first law of thermodynamics, the conservation of energy, applies to accident reconstruction. This chapter also discusses how the irreversible work done in crushing a vehicle can be quantified and related back to the pre-impact velocities of the interacting vehicles. Quantifying the various ways in which energy is dissipated in an accident to determine the pre-impact kinetic energies of the vehicles, and thus their pre-impact velocities, is known as the energy method.

In general, the application of the energy method is very simple. The total energy of a system prior to an accident is equated to the total energy of the system after the accident plus the irreversible work done during the accident interval.

$$E_1 = E_2 + U \qquad \text{(i)}$$

where

E_1 = total energy of the system, kinetic and potential, prior to the accident,

E_2 = total energy of the system, kinetic and potential, after the accident, and

U = total amount of irreversible work done during the accident, between states "1" and "2".

Prior to an accident, the interacting vehicles have a certain total amount of kinetic and potential energy, "E_1." The kinetic energy will be associated with the vehicles' velocity and mass. The potential energy will be associated with the vehicles' elevation. In the course of the accident, some of the total energy will be dissipated as irreversible work, "U." Also, some of the kinetic energy might be converted into potential energy or vice versa.

The initial kinetic energies of the interacting vehicles, and therefore their initial velocities, can be determined then if:

a. The amount of irreversible work is quantifiable.
b. The initial and end states for potential energy are known.
c. The kinetic energies possessed by the interacting vehicles after impact (or whatever) are known.

Two other items are also noteworthy in the application of energy methods. The first is that, in principle, kinetic energy can be reversibly converted into potential energy and vice versa. In practice, however, at least some of the energy in the conversion is lost due to frictional, that is, irreversible, work being done.

For example, if a car rolls downhill, its potential energy is converted into kinetic energy. However, due to rolling resistance, air resistance, tire hysteresis, and bearing friction, some of the kinetic energy is dissipated. Thus, the conversion from potential to kinetic (or vice versa) is not fully reversible. In many situations, though, the simplifying assumption of reversibility provides a reasonable model.

The second item is that in accident reconstruction, irreversible work is done by the conversion of kinetic energy. Potential energy is not directly converted into irreversible work. Potential energy must first be converted into kinetic energy before it can be dissipated as irreversible work.

For example, if the elevation of a car changes so that its potential energy is decreased, the front end of the car does not by itself become crushed. However, if the same car lowers its potential energy by rolling downhill, the car gains speed as the potential energy is converted into kinetic energy. If the car then collides with a brick wall, the front end will be crushed as its kinetic energy is converted into irreversible work.

The energy method is an extremely powerful tool for the solution of pre-impact velocities. In conjunction with the application of the momentum equations, it forms the basis for the solution of most vehicular accidents. Sometimes information missing to complete the solution by one method can be supplied by the other. Sometimes both can be used independently to solve the same problem. When both methods independently converge upon the same solution, it tends to verify the solution.

B. Some Theoretical Underpinnings

First, let us consider what is called a conservative system. A conservative system is one in which there is complete reversibility between kinetic and potential energy, and there is no irreversible work. The sum of the kinetic and potential energies of the system is the total energy.

$$E = KE + PE = \text{constant} \tag{ii}$$

where

E = the total energy of system,
KE = the kinetic energy, and
PE = the potential energy.

Taking the differential of the above gives:

$$dE = d(KE + PE) = d(KE) + d(PE) = 0 \tag{iii}$$

Now for the moment, consider just the kinetic energy term, "KE." In a fully conservative system, the kinetic energy of an object depends on its velocity, which in turn depends on its position. In most vehicular accident cases, it is practical to apply the standard Cartesian "x-y-z" orthogonal coordinate system. In functional notation then, the function for kinetic energy is given by the following:

$$KE = KE(x,y,z,u,v,w) \tag{iv}$$

where

x = displacement in the x direction,
y = displacement in the y direction,
z = displacement in the z direction (elevation),
u = dx/dt
v = dy/dt, and
w = dz/dt.

The total differential of "KE" is then as follows:

$$d(KE) = [\delta(KE)/\delta x]dx + [\delta(KE)/\delta y]dy + [\delta(KE)/\delta z]dz +$$

$$[\delta(KE)/\delta u]du + [\delta(KE)/\delta v]dv + [\delta(KE)/\delta w]dw \qquad (v)$$

From previous considerations (see Chapter 2, Section B), it is known that:

$$KE = \tfrac{1}{2}\, m[u^2 + v^2 + w^2] \qquad (vi)$$

If the gradient of the above scalar function is taken with respect to velocity, then the following vector relationship is derived:

$$[\delta(KE)/\delta u]i + [\delta(KE)/\delta v]j + [\delta(KE)/\delta w]k = \tfrac{1}{2}\, m[2u + 2v + 2w] \qquad (vii)$$

where i, j, and k are unit vector in the x, y, and z directions respectively.

If the above vector relationship is dot multiplied by the velocity vector, "V = u + v + w," then the following is obtained:

$$[\delta(KE)/\delta u]u + [\delta(KE)/\delta v]v + [\delta(KE)/\delta w]w =$$

$$\tfrac{1}{2}\, m[2u^2 + 2v^2 + 2w^2] \qquad (viii)$$

Inspection of equation (viii) finds that the right side of the equation is simply twice the kinetic energy, or "2(KE)." Thus, the following is noted:

$$2(KE) = [\delta(KE)/\delta u]u + [\delta(KE)/\delta v]v + [\delta(KE)/\delta w]w \qquad (ix)$$

If equation (ix) is differentiated, then the following additional relation is obtained:

$$2[d(KE)] = d[\delta(KE)/\delta u]u + d[\delta(KE)/\delta v]v + d[\delta(KE)/\delta w]w$$

$$+ [\delta(KE)/\delta u]du + [\delta(KE)/\delta v]dv + [\delta(KE)/\delta w]dw \qquad (x)$$

Subtracting equation (v) from equation (x) gives the following:

$$d(KE) = d[\delta(KE)/\delta u]u + d[\delta(KE)/\delta v]v + d[\delta(KE)/\delta w]w$$

$$+ [\delta(KE)/\delta u]du + [\delta(KE)/\delta v]dv + [\delta(KE)/\delta w]dw$$

$$- [\delta(KE)/\delta x]dx - [\delta(KE)/\delta y]dy - [\delta(KE)/\delta z]dz$$

$$- [\delta(KE)/\delta u]du - [\delta(KE)/\delta v]dv - [\delta(KE)/\delta w]dw \qquad \text{(xi)}$$

$$d(KE) = d[\delta(KE)/\delta u]u + d[\delta(KE)/\delta v]v + d[\delta(KE)/\delta w]w$$

$$- [\delta(KE)/\delta x]dx - [\delta(KE)/\delta y]dy - [\delta(KE)/\delta z]dz$$

Substituting "dx/dt", "dy/dt", and "dz/dt" in for "u", "v", and "w" respectively, and shifting the "dt" term to the differential operator, the following final form is obtained:

$$d(KE) = [d/dt][\delta(KE)/\delta u](dx) + [d/dt][\delta(KE)/\delta v](dy) + [d/dt][\delta(KE)/\delta w](dz) - [\delta(KE)/\delta x]dx - [\delta(KE)/\delta y]dy - [\delta(KE)/\delta z]dz$$

$$d(KE) = \{[d/dt][\delta(KE)/\delta u] - [\delta(KE)/\delta x]\}(dx) + \{[d/dt][[\delta(KE)/\delta v] - [\delta(KE)/\delta y]\}(dy) + \{[d/dt][\delta(KE)/\delta w] - [\delta(KE)/\delta z]\}(dz) \qquad \text{(xii)}$$

By the above manipulations, the terms "du", "dv", and "dw" which were present in equation (v), have been eliminated.

Consider now the term for potential energy, "PE". The potential energy of a conservative system is simply a function of position. In fact, if the potential energy under consideration is simply that of gravity, then "PE" is simply a function of one coordinate, "z".

$$PE = PE(z) \qquad \text{(xiii)}$$

The differential of "PE" is then:

$$d(PE) = [\delta(PE)/\delta z](dz) \qquad \text{(xiv)}$$

Combining equation (iii) with equations (xii) and (xiv) gives the following:

$$dE = d(KE) + d(PE) = \{[d/dt][\delta(KE)/\delta u] - [\delta(KE)/\delta x]\}(dx) + \{[d/dt][\delta(KE)/\delta v] - [\delta(KE)/\delta y]\}(dy) + \{[d/dt][\delta(KE)/\delta w] - [\delta(KE)/\delta z] + [\delta(PE)/\delta z]\}(dz) = 0 \qquad \text{(xv)}$$

Since the "x–y–z" coordinates are independent of one another, then the above can be broken apart into the following independent coordinate equations:

$$[d/dt][\delta(KE)/\delta u] - [\delta(KE)/\delta x] = 0 \qquad \text{(xvi)}$$

$$[d/dt][\delta(KE)/\delta v] - [\delta(KE)/\delta y] = 0$$

$$[d/dt][\delta(KE)/\delta w] - [\delta(KE)/\delta z] + [\delta(PE)/\delta z] = 0$$

Equations (xvi) are generally called *Lagrange's equations* for a conservative system. In many texts, the following substitution is usually made:

$$L = KE - PE \quad \text{or} \quad KE = L + PE \qquad \text{(xvii)}$$

where L is called the *Lagrangian term.*

The substitution of "KE = L + PE" changes equations (xvi) to the following:

$$[d/dt][\delta(L)/\delta u] - [\delta(L)/\delta x] = 0 \qquad \text{(xviii)}$$

$$[d/dt][\delta(L)/\delta v] - [\delta(L)/\delta y] = 0$$

$$[d/dt][\delta(L)/\delta w] - [\delta(L)/\delta z] = 0$$

Note: $\delta(PE)/\delta u = 0$
$\delta(PE)/\delta v = 0$
$\delta(PE)/\delta w = 0$
and $\delta(PE)/\delta x = 0$
$\delta(PE)/\delta y = 0$

It is worth noting at this point that the term "$[\delta(L)/\delta u]$" is simply the momentum of the system.

$[\delta(L)/\delta u] = \delta[KE - PE]\delta u = mu$ remembering that $\delta(PE)/\delta u = 0$.

In fact, the gradient of the kinetic energy term, as was taken in equation (vii), is simply the vector representation of the kinetic energy term:

$$\text{grad}(KE) = [\delta(KE)/\delta u]i+[\delta(KE)/\delta v]j+[\delta(KE)/\delta w]k = m[u + v + w].$$

In a sense, the momentum equations already discussed in Chapter 9 are seen to be imbedded in the above energy equations of a conservative system.

The Lagrangian equations in equation (xviii) can be further generalized by the introduction of generalized coordinates. In this way, equations (xviii) can be applied to any orthogonal coordinate system. While for nearly all vehicular accidents the Cartesian coordinate system is convenient, occasionally the cylindrical coordinate system may be useful. In such a system,

$$x = r \cos \theta,\ y = r \sin \theta,\ r = (x2 + y2)^{1/2},\ \text{and}\ z = z.$$

With appropriate substitutions, the Cartesian oriented Lagrangian equations in equation (xviii) can be converted into a cylindrical coordinate system.

At this point, equations (xviii) are merely an elegant way of describing the "sloshing" of energy from kinetic to potential and vice versa in a fully conservative system. In celestial mechanics, the Lagrangian equations are a cornerstone with respect to central forces and orbital motion. Unfortunately, in vehicular accident reconstruction, about the only situation to which equations (xviii) apply are vehicles coasting up or down hills or perhaps a car in orbit. To make the Lagrangian equations useful for vehicular accident reconstruction, it is necessary to introduce irreversible work into the equation. This can be done as follows:

$$d(KE + PE) = dU \qquad \text{(xix)}$$

where U = irreversible work.

Work is defined as force applied through a distance, or "dU = F · dx." The differential "dU" is the dot product of a force, "F," and an infinitesimal distance "dx" through which the force is applied.

In the above, the differential "dU" is the result of one specific force being applied in a correspondingly specific direction. In a vehicular accident, however, there can be many ways in which irreversible work can be done (e.g. skidding, crush, rolling friction). Because of this, the above term for irreversible work is expanded to include many such forces and their corresponding coordinate directions of application.

$$dU = \sum [F_i \cdot dx + G_i \cdot dy + H_i \cdot dz] \qquad \text{(xx)}$$

where F, G, and H = forces in the respective x, y, and z directions.

If equation (xx) is substituted into the right side of equation (xix), and equation (xv) is substituted into the left side of equation (xix), and the results are separated into respective coordinate equations, then the following is obtained:

$$\{[d/dt][\delta(KE)/\delta u] - [\delta(KE)/\delta x]\}(dx) = \sum [F_i \cdot dx] \qquad \text{(xxi)}$$

$$\{[d/dt][\delta(KE)/\delta v] - [\delta(KE)/\delta y]\}(dy) = \sum [G_i \cdot dy]$$

$$\{[d/dt][\delta(KE)/\delta w] - [\delta(KE)/\delta z] + [\delta(PE)/\delta z]\}(dz) = \sum [H_i \cdot dz]$$

The reader will note that the Lagrangian term "L" was dropped in favor of using "KE's" and "PE's". This is because the potential energy term is only important in the "z" direction coordinate.

Finally, the integration of equations (xxi) gives the following:

$$\int_{x_1}^{x_2} \{[d/dt][\delta(KE)/\delta u]-[\delta(KE)/\delta x]\}(dx) = \int_{x_1}^{x_2} [\sum F_i \cdot dx] \qquad \text{(xxii)}$$

$$\int_{y_1}^{y_2} \{[d/dt][\delta(KE)/\delta v]-[\delta(KE)/\delta y]\}(dy)] = \int_{y_1}^{y_2} [\sum G_i \cdot dy]$$

$$\int_{z_1}^{z_2} \{[d/dt][\delta(KE/\delta w]-[\delta(KE)/\delta z]+[\delta(PE)/\delta z]\}(dz) = \int_{z_1}^{z_2} [\sum H_i \cdot dz]$$

where the subscripts "1" and "2" refer to the pre-accident and post-accident positions.

Inspection of equations (xxii) finds that:

- all the terms on the right hand are irreversible work,
- all the terms on the left are related to the kinetic or potential energy of the object,
- the first term on the left side is a work term, i.e., force x distance, where an external force is being applied on the item,
- the second term on the left side is the change in kinetic energy from position "1" to position "2", and
- irreversible work done by the object is done in the direction of travel of the object (in that coordinate frame).

If the usual terms for kinetic and potential energy are substituted into equations (xxii), the following is obtained:

$$F_{x,ave}(x_2 - x_1) - \tfrac{1}{2}(m)(u_2^2 - u_1^2) = U_x \qquad \text{(xxiii)}$$

$$F_{y,ave}(y_2 - y_1) - \tfrac{1}{2}(m)(v_2^2 - v_1^2) = U_y$$

$$F_{z,ave}(z_2 - z_1) - \tfrac{1}{2}(m)(w_2^2 - w_1^2) + g(z_2 - z_1) = U_z$$

In general, when no net external reversible forces are acting upon a vehicle, the first terms on the left hand side of equations (xxiii) involving force are zero. Since all of the terms in equation (xxiii) are scalar, then the following is also true:

$$\tfrac{1}{2}(m)[(u_1^2 - u_2^2) + (v_1^2 - v_2^2) + (w_1^2 - w_2^2) + 2g(z_2 - z_1)] =$$

$$U_x + U_y + U_z \qquad \text{(xxiv)}$$

Of course, the above is the equation for only one object, or one vehicle. For two vehicles colliding or nearly colliding with one another, the following holds:

$$\tfrac{1}{2}(m_A)[(u_{A1}^2 - u_{A2}^2) + (v_{A1}^2 - v_{A2}^2) + (w_{A1}^2 - w_{A2}^2) + 2g(z_{A2} - z_{A1})]$$

$$+ \tfrac{1}{2}(m_B)[(u_{B1}^2 - u_{B2}^2) + (v_{B1}^2 - v_{B2}^2) + (w_{B1}^2 - w_{B2}^2) + 2g(z_{B2} - z_{B1})]$$

$$= U_{Ax} + U_{Ay} + U_{Az} + U_{Bx} + U_{By} + U_{Bz} = U_{Total} \qquad \text{(xxv)}$$

Equation (xxv) represents the general energy equations for a two-car accident in Cartesian coordinates. Equation (xxv) can also be represented in cylindrical coordinates with the following substitutions:

$$R_1^2 = [u_1^2 + v_1^2 + w_1^2] \qquad \text{(xxvi)}$$

$$R_2^2 = [u_2^2 + v_2^2 + w_2^2]$$

Rearranging terms and substituting gives the following:

$$\tfrac{1}{2}(m_A)[R_{A1}^2 - R_{A2}^2 + 2g(z_{A2} - z_{A1})] \qquad \text{(xxvii)}$$

$$+ \tfrac{1}{2}(m_B)[R_{B1}^2 - R_{B2}^2 + 2g(z_{B2} - z_{B1})] = U_{AR} + U_{BR} = U_{Total}$$

Using cylindrical coordinates removes the "x-y" related velocity components in equations (xxv) and simplifies the form of the energy equation. In essence, since equation (xxvii) is scalar, it is only necessary to work with the absolute magnitudes of the velocities, and the absolute magnitudes of the irreversible work terms.

The above is very important. In the basic energy method, it is not necessary to know from which direction the vehicles came, in which direction they left, or in which direction the irreversible work was done. It is only important to know the absolute values of the various energy terms. In short, the energy method is more or less like balancing sums in an accountant's ledger.

Since the terms on the left hand of equations (xxv) or (xxvii) have already been discussed in Chapters 2 and 3 at some length, all that remains is to figure out how to quantify the irreversible work of the right hand terms.

C. General Types of Irreversible Work

In vehicular accidents, the most common forms of irreversible work are as follows:

a. Braking and skidding,
b. Sheet metal crush of the vehicle(s),
c. Rollovers,
d. Crush or impact damage to non-vehicles, such as telephone poles, bridge railings, sign posts, etc.

If a person wants to split hairs, items "b" and "c" above are both types of sheet metal damage. However, the mechanics in a rollover are sufficiently different from typical impact crush to warrant its consideration separately. A fifth, miscellaneous category can be added which lumps together less common types of irreversible work. Some of these miscellaneous types of irreversible work include:

a. Driving into a deep pool of water,
b. Driving into saturated sand, mud, deep snow or a freshly disked field,
c. Driving into a row of haystacks, and
d. Allowing the air resistance, rolling friction, and engine drag to slow the vehicle down to a stop.

In general, the above miscellaneous irreversible work items involve slow-acting braking that may cause only limited damage to the vehicle itself. These cases can be difficult to quantify directly and may require special case-by-case consideration. Often, such cases require on-site experimentation to determine the pertinent parameters.

When a vehicle is being extracted from mud, sand, or whatever, sometimes it is useful to place a load cell in the tow cable to measure the tow draw. By comparing this to the tow draw of the vehicle on level, solid pavement, the frictional resistance of the medium can be measured.

However, the above may not be sufficient for a pool of water. The resistance measured when a vehicle is slowly towed from a pool of water may not be the same as that experienced by the vehicle traveling through it at driving speeds. In general, the frictional resistance of water is a function of the forward velocity of the vehicle in much the same way as air resistance. A further complication may occur if the vehicle was traveling fast enough to hydroplane over the surface of the water before sinking. And of course, there is the obvious complication of dealing with water; water does not normally exhibit skid marks to mark a vehicle's path.

Fortunately (or unfortunately for the driver), when a vehicle is driven off a dock, pier, or similar, it usually has a sufficient angle of attack with the water to "bite" or "catch" the water, rather than to skip or hydroplane over it. The point where the vehicle sinks will often be the terminus of the trajectory from the dock, pier, or whatever. In such cases, the problem then can be solved as a simple trajectory, as discussed in Chapter 3.

With respect to air resistance, rolling resistance and engine drag, these topics have already been discussed in Chapters 3 and 4. The braking and skidding category of irreversible work has been discussed at some length in Chapter 2. Thus, the remaining items to consider are rollovers and crush damage.

D. Rollovers

A side rollover is one in which the vehicle initially rotates about the line formed by either the right side or left side wheels. Kinetic energy from the motion of the car is converted into work, which then lifts the center of gravity of the vehicle by rotation. When the vehicle is so lifted, its potential energy is increased. As the vehicle rotates or rolls over through 360 degrees, it will pass through four balance or instability points. Each of these instability points is a position where the local potential energy of the vehicle is at a maximum.

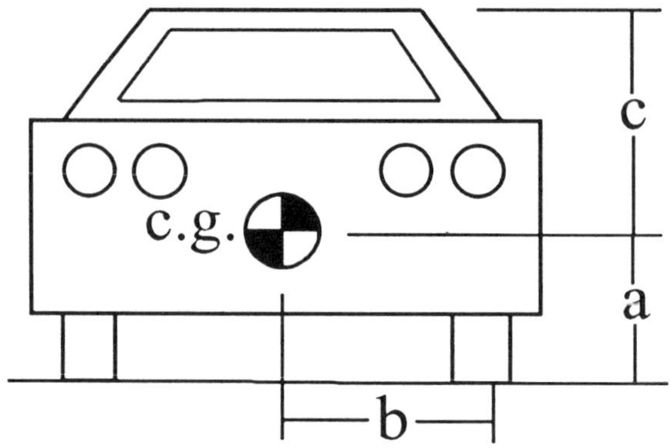

Figure 1. C.G. (Center of Gravity) from end of vehicle

If the conversion of kinetic energy to work is sufficient to lift the vehicle's center of gravity higher than the elevation of the first localized potential maximum, the vehicle will continue to roll over. If the conversion of kinetic energy is insufficient to lift the center of gravity above the first localized potential maximum, the vehicle will simply lift up on its side, and then fall back down onto its wheels.

Figure 1 shows the front view of a vehicle with the center of gravity located "a" distance from the pavement, and "b" distance from the outer edge of the tires. The first instability point, or localized potential maximum, of a vehicle experiencing a side rollover is given as follows:

$$h_1 = [a^2 + b^2]^{1/2} \qquad \text{(xxviii)}$$

where h_1 = first localized maximum potential during side rollover.

It should be noted that "h_1" as given above is not exact. During tipping of the vehicle, the load carried by the tires and suspension on the side not tipped will double, causing some additional downward displacement or "squat." If the tipping is rapid, the downward displacement may be quite negligible since the suspension system may be damped such that it does not have time to respond. In any case, if needed, a "squat" version of equation (xxviii) is given as follows:

$$h_1 = [a^2 + b^2]^{1/2} - (W/2k) \qquad \text{(xxix)}$$

where

$W/2$ = extra load being carried by the side wheels during tipping,
k = linear spring constant of the suspension system where
$-(\Delta z)k$ = applied force.

For purposes of this section, equation (xxviii) gives a sufficient approximation for "h_1".

The angle at which the vehicle must be tipped to be balanced at the first instability point is given by the following:

$$\alpha = \arctan (b/a) \qquad \text{(xxx)}$$

The angle "α" is usually given in degrees.

In a typical car where a = 24 inches, b = 33 inches, and c = 30 inches, "α" is about 54 degrees, and "h_1" is about 40.8 inches.

A second localized potential maximum occurs when the vehicle rolls from the 90-degree position (on its side) to its 180-degree position (on its top). If the vehicle is more or less box shaped, the second localized potential maximum is given by the following:

$$h_2 = [b^2 + c^2]^{1/2} \qquad \text{(xxxi)}$$

The position of the second localized potential maximum is given by the following:

$$\beta = 90 + \arctan (c/b) \qquad \text{(xxxii)}$$

If the same dimensions as noted above are substituted into equations (xxxi) and (xxxii), it is found that "h_2" is 44.6 inches, and "β" is 132 degrees.

There are also a third and a fourth localized potential maximums that are symmetric to the first two. Using the same dimensions as before, the third point is located at an angle of 228 degrees with an "h_3" of 44.6 inches, and the fourth point is located at an angle of 306 degrees with an "h_4" of 40.8 inches.

Based on the same vehicle dimensions as given above, the following table shows the relative elevations of the center of gravity as the vehicle rolls over sideways through 360 degrees. It is assumed that no significant crush occurs during rollover to change the initial dimensions of the vehicle.

Figure 2. Rollover damage to roof of Thunderbird.

Table 1. Elevation of Center of Gravity - Rollover

Angular position in degrees	elevation of c.g. in inches	remarks
0	24	upright
54	41	1st max.elev.of c.g.
90	33	on side
132	45	2nd max. elev. of c.g.
180	30	upside down
228	45	3rd max.elev.of c.g.
270	33	on other side
306	41	4th max.elev.of c.g.
360	24	upright

In essence, if the various values for the center of gravity elevations listed in table 1 are plotted against the corresponding angles of rotation of the rollover, a plot of the potential energy versus rollover angle is generated. Figure 3 is such a plot.

Inspection of the graph in Figure 3 finds several interesting features. First of all, the position of the car when it is upright is the position which minimizes the potential energy. This is the most stable position of the vehicle. The second most stable position is when the car is on its roof. In order for the roof to be the most stable position, it would have to be crushed more than 6 inches.

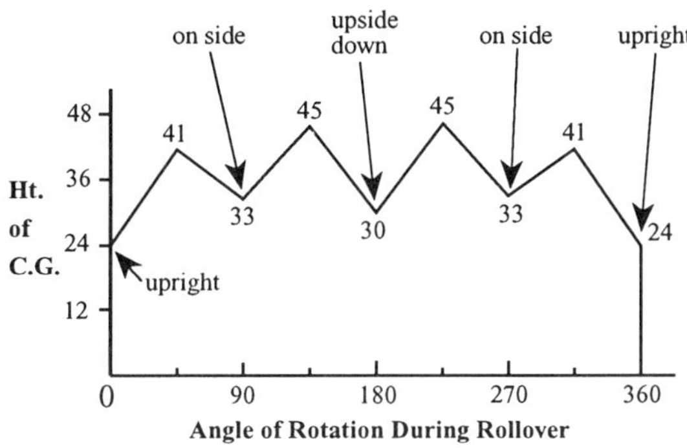

Figure 3. Plot of C.G. versus Roll Over Angle

During rollover, the center of gravity of the vehicle has four "peaks" of energy to climb and four "hills" of energy to slide down. The first "peak" is a change of 17 inches. After the first "peak", there is an 8-inch "slide" to the side or 90 degree rest position. The second peak is a change of 12 inches from the 90 degree position, and a total change of 21 inches from the initial position of 0 degrees. After the second "peak," there is a "slide" of 15 inches to the 180 degree position, i.e., upside down.

From experience with playing with old heaps in junk yards, the following "rules" with respect to the example vehicle appear to apply to rollovers:

1. If there is just enough work performed to roll the car over the first "peak", the car will then come to rest on its side. It takes 17 inches of energy to accomplish this. Eight inches of energy will be lost in crush, damping, and friction. The other nine inches of energy goes to increasing the overall potential energy of the vehicle.
2. If there is just enough work performed to roll the car onto its roof, another twelve inches of energy will be needed to go from the side position to the roof position. This is a total of 29 inches of energy expended so far.
3. To continue rolling over from the upside down position onto its other side, another 15 inches of energy is needed to clear the next "peak." This is a total of 44 inches of energy expended so far.
4. To finish the roll and land back upright, another 8 inches of energy is needed to clear the last "peak." This then is a total of about 52 inches of energy to complete a 360-degree rollover.

The above scenario assumes that the vehicle's energy on the "downhill" side of the peaks in figure 2 is dissipated along the way by friction, crush, damping, etc. For most instances, this is a reasonable approximation when the vehicle rolls over more or less in contact with the ground.

Generalizing from the above, the following equations indicate what the expenditure of energy will be for the various rest positions after a roll over.

Case I. Near rollover. The vehicle lifts up, but does not clear the first instability point, and drops back down onto its wheels.

$$U < W(h_1 - a) \qquad \text{(xxxiii)}$$

$$U < W([a^2 + b^2]^{1/2} - a)$$

where

U = work expended in roll over action, and

W = weight of the vehicle.

Case II. Rollover which ends up on side position.

$$U = W(h_1 - a) \qquad \text{(xxxiv)}$$

Case III. Rollover which ends up on roof.

$$U = W([b^2 + c^2]^{1/2} - b + [a^2 + b^2]^{1/2} - a) \qquad \text{(xxxv)}$$

$$U = W(h_1 - a + h_2 - b)$$

Case IV. Rollover which ends up on the other side of the vehicle.

$$U = W([b^2 + c^2]^{1/2} - b + [a^2 + b^2]^{1/2} - a + [b^2 + c^2]^{1/2} - c)$$

$$U = W(h_1 - a + 2h_2 - b - c) \qquad \text{(xxxvi)}$$

Case V. Rollover which ends up upright again.

$$U = W([b^2 + c^2]^{1/2} - 2b + 2[a^2 + b^2]^{1/2} - a + [b^2 + c^2]^{1/2} - c)$$

$$U = W(2h_1 - a + 2h_2 - 2b - c) \qquad \text{(xxxvii)}$$

Of course, it should be noted that the above equations have been developed for a car:

1. That is more or less a solid rectangle,
2. In which no significant crush has occurred to grossly change initial dimensions,
3. In which the "squat" due to load transfer to one side of the vehicle during the initial roll over is negligible.

The above equations would not exactly work for pickup trucks, since trucks might "teeter" front to back when they are upside down. Similarly, the tractors in tractor-trailer rigs would likely not turn upside down as a car would. They would more likely roll over cocked, with ground contact simultaneously occurring at the front edge of the hood, and the top edge of the roof. However, the principles demonstrated for developing equations for a solid rectangular car are the same for other shapes of vehicles.

If it is assumed that the same example vehicle is used as before, and that the sample vehicle has a weight of 2,500 lbs, then the minimum velocity at which a rollover could occur is calculated as follows:

$$\tfrac{1}{2}\, mv^2 = W([a^2 + b^2]^{1/2} - a) \qquad \text{(xxxviii)}$$

$$v = \{2g([a^2 + b^2]^{1/2} - a)\}^{1/2}$$

$$v = 9.5 \text{ ft/sec.}$$

Note that the weight of the vehicle is partially cancelled out by the mass term on the left side, so that the equation ends up using "g." Equation (xxxviii) is independent of the vehicle's weight.

The above solution found in equation (xxxviii) does not mean that a vehicle will turn over whenever it is traveling sideways at 9.5 ft/sec. However, if it were traveling sideways at 9.5 ft/sec and its side wheels caught a curb or something that momentarily held the wheels immobile, then rollover could occur. The velocity term in equation (xxxviii) is the sideways or transverse speed of the center of gravity when the wheels are simultaneously prevented from having transverse motion.

If the same things are assumed about the example vehicle as before, but the rollover continues through 360 degrees until the vehicle is upright again, then the minimum velocity at which such a rollover could occur is calculated as follows:

$$\frac{1}{2} mv^2 = W(2h_1 - a + 2h_2 - 2b - c) \quad \text{(xxxix)}$$

$$v = 16.5 \text{ ft/sec.}$$

The above calculations in equations (xxxviii) and (xxxix) demonstrate that it really does not take a lot of speed to have enough energy to roll a vehicle over. In other words, a sideways rollover does not expend a lot of energy. For example, the energy required for the 360-degree rollover of equation (xxxix) is equivalent to just a 6 ft. long skid if the coefficient of friction is 0.7.

E. Flips

A "flip" happens when a car or other type of vehicle rotates end over end. Usually, the rear end of the vehicle rises and rotates about the front wheels. Most of the time when a vehicle flips over, it is the rear end that rises up because most vehicles flip over when they are moving in a forward direction.

Like the rollover, a flip occurs when the center of gravity of the vehicle is traveling with a certain velocity while the front two (or possibly back two) wheels are momentarily motionless. Such a situation can occur if:

1. The vehicle hits a low, retaining wall or embankment.
2. The vehicle drives off the road such that the front end of the car drops down and "catches" in the dirt.
3. The vehicle drives into a high curb.
4. The vehicle drives into a mound of gravel.

In each of the above cases, the front end of the car is momentarily stopped, while the rest of the car is still able to move by rotating about the front wheels. This occurs because the car's center of gravity is located above the point of force which stops the car, as shown in figure 3. If the center of gravity were in line with the point of force, the car would not rise up and rotate. If the center of gravity were below the point of force, the vehicle would "squat" during the impact as it tries to rotate into the ground. (In head-on collisions or similar, it is not unusual for one or both cars to "squat" enough to cause some of the tubeless tires to lose their seal and deflate.)

A flip is much simpler to calculate than a rollover. While a rollover has four points of instability and four points of local stability (upright, right side, upside down, and left side), a flip has only two points of instability and two points of local stability. The two points of instability are the front and the back, that is, being stood on either end. The two points of stability are upright and upside down. All these points, that is both the stable and unstable points, are basically spaced 90 degrees from one another.

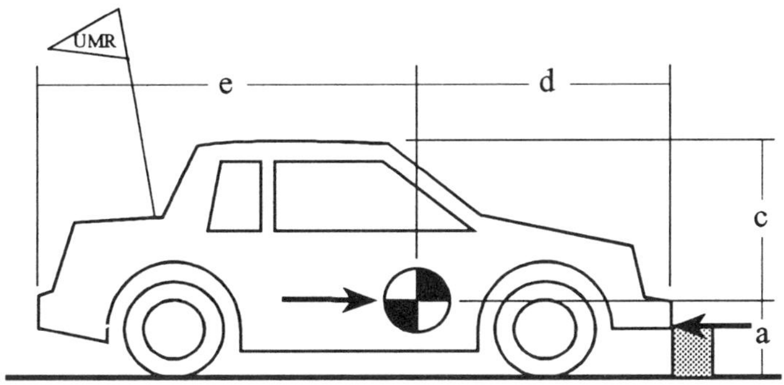

Figure 4. Position of C.G. from side of vehicle

While it may be arguable that in a vehicle with a squared off front and rear there are theoretically four points of instability and four points of stability to a flip, there is no practical significance to considering all four points. The first two instability points, or potential energy "peaks," are so close to one another, and the intervening potential energy "valley" between them is so slight, that there is no practical significance to separating the three points. This is also true of the second two instability points and the intervening potential energy "valley."

The above should be intuitively obvious. Vehicles just don't regularly flip over and come to rest poised like balancing rocks standing on their ends. It is very difficult to stand a vehicle on either its front end or rear end and have it stay put. While this possibly could be done with a delicate "touch" and a very squarish vehicle, such a delicate "touch" does not occur in vehicular accidents. Vehicles in accident situations usually just smash around like rogue elephants.

As before with rollovers, the energy needed to flip a vehicle over is basically a function of the work it takes to lift its center of gravity. Assume that the same sample car as considered before is used and it has the same center of gravity, that is, a = 24 in., b = 33 in., and c = 30 in. Further, assume that as shown in figure 3, the other dimensions locating the center of gravity from the side are d = 75 in., and e = 115 in. Then a table similar to Table 1 can be made which shows the variation in center of gravity elevation as a vehicle flips forward.

Table 2. Elevation of Center of Gravity - Flip

angular position in degrees	elev. of c.g. in inches	remarks
0	24	upright
90	75	standing on front end
180	30	upside down
270	115	standing on rear end
360	24	upright

As in the example vehicle, most cars have their center of gravity significantly closer to the front end of the car than to the rear. This means that it is significantly easier to flip a car over the front end than over the rear end. If the values in table 2 were to be plotted like those in table 1, the "peaks" would occur at 90 and 270 degrees, and the "valleys" would occur at 0 and 180 degrees.

While table 2 shows the elevation of the center of gravity through a 360-degree flip, it is very unusual for a car to flip through one full rotation. In many cases, a car will first flip over onto its top, and if there is enough remaining kinetic energy, it may then roll over rather than continue with the flip. There is a tendency for rolling over even when there is sufficient energy available to continue the flip. The underlying reasons for this behavior are rooted in rotational dynamics and deserve a brief look even if they do not exactly relate to the topic at hand.

A similar rotational effect occurs with a book that is significantly longer than it is wide. If a person tries to "roll" the book, that is rotate it about a line parallel to its binding, the long axis, the book will spin stably about the axis. If a person then attempts to rotate the book about a line parallel with its top edge across, the short axis, the book may begin to tumble or wobble as it rotates after about 180 degrees of spin. It is demonstrably harder to make the book rotate cleanly about its short axis than its long axis.

This effect is a consequence of *Euler's Equations of Motion*, and the mass moments of inertia in the three principal axes. Euler's Equations of Motion, shown below, relate torque, angular velocity, angular acceleration, and the mass moments of inertia. They are applied to items like tops, gyroscopes and other objects which rotate in two or three dimensions. Without showing the derivations, in a body that has symmetry about each of its three principal axes, the following equations hold:

$$T_x = I_{xx}\alpha_x + (I_{zz} - I_{yy})\omega_z\omega_y$$

$$T_y = I_{yy}\alpha_y + (I_{xx} - I_{zz})\omega_x\omega_z$$

$$T_z = I_{zz}\alpha_z + (I_{yy} - I_{xx})\omega_y\omega_x \qquad \text{(xl)}$$

where

T_i = moment or torque about the "i" axis,
I_{ii} = mass moment of inertia about the "i-i" axis,
ω_i = angular velocity about the "i" axis, and
α_i = $d(\omega_i)/dt$, angular acceleration about the "i" axis.

(Note: An excellent derivation of Euler's Equations of Motion can be found in the text "Dynamics," by Pestel and Thomson, McGraw-Hill, 1968, pages 349–353.)

Two items are apparent upon inspection of the above equations. First, if the mass moments of inertia are all equal, as they would be in a sphere, then all three equations simply become "$I_{ii}\ \alpha_i = T_i$." When all three mass moments of inertia are different from each other, it is possible for linkage to occur among the rotational motions of the three axes. This linkage causes rotational motions to behave much differently than linear motions. This is the reason why a top precesses when it leans over while spinning, and is one of the reasons why it is easier to balance a moving bicycle than one that is stopped.

If the coordinates of the vehicle are oriented such that "x–x" is along a line from the right side to the left side of the vehicle, "y–y" is along a line from the front to the rear, and "z–z" is along a line from the bottom to the top, in general it is found that "I_{xx}" and "I_{zz}" are equal to one another. This is because the distribution of mass from the axis of rotation is about the same in both cases for most types of vehicles.

For most vehicles, the least mass moment of inertia is about the "y–y" axis, and it is usually significantly less than the other two mass moments of inertia along the principle axes. In the "I_{yy}" term, the line of rotation goes through more of the mass, which "zeroes" out that portion of the mass in the moment of inertia calculation. Also, the longest distance from the axis is perhaps only 1/3 or 1/4 that of the other axes.

Thus, it is easier to rotate a car about its short axis than about its long axis. A greater rotational response occurs when a torque is applied to the "I_{yy}" term than would occur if applied to the other two moments of inertia.

By applying the above general findings to equations (xl), the following simplified Euler's Equations of Motion are derived which apply to many types of automobiles:

$$T_x = I_{xx}\alpha_x + (I_{zz})\omega_z\omega_y \qquad \text{(xli)}$$

$$T_y = I_{yy}\alpha_y$$

$$T_z = I_{zz}\alpha_z - (I_{xx})\omega_y\omega_x$$

where

$(I_{xx} - I_{zz}) \sim 0$, and

$I_{yy} << I_{xx}$ and I_{zz}.

Inspection of the above equations shows that torque around the "y–y" axis is not interlinked to motion about the other axes. However, torque in either the "x–x" or "z–z" axis is interlinked to motions in the other axes.

In a rollover, rotation occurs about the "y–y" axis due to the application of a torque at the bottom of the tires on one side of the vehicle. The influence of small angular velocity rotations around the other two axes upon the "y" torque is nullified by the "$(I_{xx} - I_{zz})$" term, which is zero. However, because it is a rollover action, rotation in the "x–x" axis is zero anyway. This is because prior to the rollover, the vehicle is in contact with the ground which constrains any "x–x" axis rotation. Thus, in a rollover, rotation can typically occur only in the "y–y" and "z–z" axes. Angular velocities in both of these axes could cause a torque to be developed in the "x" direction if the angular velocity in the "z–z" axis were significant. But this occurs infrequently. For these reasons, rollovers don't usually turn into flips unless the ground helps out.

In a flip, where rotation is about the "x–x" axis, the term "$(I_{zz} - I_{yy})$" is not zero. The term "$(I_{zz} - I_{yy})$" is the difference between the maximum and the minimum mass moments of inertia of the vehicle. Since "I_{yy}" is usually very small as compared to "I_{zz}", the "I_{yy}" term is neglected for simplicity. Also, there is no angular velocity about the "y–y" axis because the vehicle has been restrained from rotation in that direction by contact with the ground.

Thus, a vehicle can go into a forward flip with some previous rotation about the "z–z" axis. This could cause the vehicle to continue yawing during the flip, and land upside down in a partially or fully sideways position. Thus, if there is sufficient energy available, a vehicle could flip 180 degrees onto its top, and then go into a roll over.

In a spin, rotation is about the "z–z" axis (yaw). A vehicle that is yawing will normally not have rotations in the other two directions, since the vehicle is constrained by the ground.
The "ω_x" and "ω_y" terms are zero. This is why in Chapter 9, where spin due to impact was discussed, Euler's Equations of Motion were not brought up.

Thus, except for unusual circumstance, only flips require the application of Euler's Equations of Motion.

Returning to the main topic of this section, the minimum kinetic energy to cause a forward flip to occur, so that the car lands on its top, is given by the following:

$$\tfrac{1}{2} mv^2 = W(d - a) \qquad \text{(xlii)}$$

where

m = mass of the vehicle,
W = weight of the vehicle,
v = velocity of the vehicle, and
d and a are the coordinates of the center of gravity as shown in figures 1 and 3.

Solving for the velocity term in equation (xli) gives the following:

$$v = (2g[d - a])^{1/2} \qquad \text{(xliii)}$$

where v = minimum velocity to cause a forward flip.

For the example car, substituting in the dimensions we find that v = 16.54 ft/sec. Comparing this with the velocity required to initiate rollover we find that a forward flip requires about 74% more velocity to initiate than a rollover. In terms of energy, it needs about three times more energy to initiate a flip than it does to initiate a roll. This explains, in part, why rollovers are more common than flips.

In a forward flip, the vehicle will sometimes land on its top, the top will crush and crumple, and the motion of the vehicle will stop. However, because it takes so much more energy to flip a vehicle than roll it, there is more energy available on the "downhill" side from the first instability point. If the roof of the vehicle is rigid and elastic, it is possible that it can bounce sufficiently for the vehicle to then roll over onto its side. Whether it lands on its top or on its side, the energy expended is still the same.

If the vehicle continues to roll after landing on its side, the rest of the roll is handled as discussed in the previous section on rollovers.

If, however, the vehicle does not roll over but continues to flip over through 360 degrees, then the following holds:

$$\tfrac{1}{2}\,mv^2 = W(d - a + e - c) \qquad \text{(xliv)}$$

Solving the above gives:

$$v = (2g[d - a + e - c])^{1/2} \qquad \text{(xlv)}$$

Again substituting in the dimensions from the example vehicle we find that the minimum speed to accomplish a 360-degree forward flip is 27.0 ft/sec. This is 64% more velocity than is required minimally to cause a 360-degree rollover. In terms of energy, it requires about 2.7 times more energy to cause a 360-degree flip than a 360-degree rollover.

However, whether it be a flip or a rollover, it is noteworthy that the minimum vehicular velocity required to cause either is low. Under the right circumstances, a vehicle traveling at even relatively slow speeds has enough velocity to either flip or roll over.

F. Modeling Vehicular Crush

Several types of conceptual models have been used to represent a vehicle which has sustained crush damage. Some computer programs that analyze vehicular collisions have modeled a vehicle like an elastic ball. In such a model, the collision is considered elastic, and the spring stiffness is the same from any direction of impact. This "billiard ball" model works reasonably well when the collision impulse is in the elastic response range of the vehicle, but quickly becomes inadequate when a high degree of plastic deformation is involved. As noted in Chapter 9, significant plastic deformation occurs at all but very modest vehicular speeds. However, the model does have some usefulness in glancing impacts at moderate to high speeds where the central impact vector component is very small as compared to the parallel impact vector.

An alternative conceptual model used in many analyses is to assume that the deformation response of the vehicle is linear. This approach is then used to develop delta-V versus crush depth relations. This is the conceptual model used in EDCRASH, a popular computer program used in analyzing collisions. In essence, this model assumes that the rate of crush deflection at the beginning of contact between the vehicles (or whatever) is equal to the closing velocity between the vehicles.

$$dx/dt = v_{A1} - v_{B1} = \Delta v \qquad \text{(xlvi)}$$

where

x = crush displacement,
v_{A1} = pre-impact speed of vehicle "A" and
v_{B1} = pre-impact speed of vehicle "B."

A crush constant "k" is then assigned to each vehicle so that the energies dissipated in crush by the respective vehicles are:

$$E_A = \tfrac{1}{2} k_A x_A^2 \text{ and } E_B = \tfrac{1}{2} k_B x_B^2 \qquad \text{(xlvii)}$$

In this model, the crush response of the vehicle is basically treated like a linear spring. The energy dissipated in crush is given the same mathematical form as the stored energy in a compressed elastic spring that obeys Hooke's Law.

Following the above model through, the function for energy consumed by crush in a vehicle is then:

$$E = (a^2/2b) + ac + b(c^2/2) \qquad \text{(xlviii)}$$

where

a and b = arbitrary constants associated with each vehicle's relative stiffness, and
c = average crush depth across the front of the vehicle, assuming that the collision was more or less across the front.

Because of the linear assumption of force versus crush depth, several problems have been reported with the above model. At low speeds the delta-V parameter is often underestimated. At high delta-V speeds of more than about 50 mph, it overestimates delta-V (see: "*An Overview of the Way EDCRASH Computes Delta-V*" by Day and Hargens, SAE paper 870045, page 192).

The model used in this text assumes that a vehicle is like a reinforced thin-walled enclosed shell, or a thin-walled column. This model appears to more closely resemble what actually occurs in a collision than the various "spring" models.

In a thin-walled column, when compressive load is applied at low levels, the column responds elastically in the same direction as the compression. As the load is increased, the column may undergo elastic buckling, where the shape of the column distorts in directions other than that of the applied force. However, in elastic buckling, the column is still able to snap back if the load is released. When the load is then increased further, inelastic buckling occurs. The column collapses by folding into itself somewhat like an accordion, and the deformation becomes permanent.

In most structural engineering or mechanics of materials courses at the undergraduate level, failure of a structural component is considered to have occurred when elastic buckling begins, and possibly even before then. Too much dimensional distortion in the elastic regime may render a part or component useless. The goal of good design is to avoid reaching that level of loading which causes excessive dimensional distortion or rupture of the part. Thus, while there is considerable study of what occurs below the buckling point, there is often scant attention paid to what occurs after buckling has occurred.

G. Post Buckling Behavior of Columns

The analysis of buckled elastic columns, which is sometimes known as the *problem of the elastica*, was first done by Euler more than 200 years ago. While the derivations are somewhat beyond the scope of this text, the results are important. In a simple column where each end is free to rotate and a load is applied axially at one end, the column will buckle elastically at the Euler buckling load, which is usually given as:

$$F_{cr} = K(\pi^2 EI)/L^2 \qquad \text{(xlix)}$$

where

F_{cr} = force required for elastic buckling, i.e., critical load,
E = modulus of elasticity,
K = constant for the end conditions of the column, which in this case is 1,
I = areal moment of inertia of cross section of column, and
L = length of column.

When both ends of the column are fixed, the value of "K" is 2. Thus, a column with fixed or clamped ends will have a Euler buckling load twice that of one that has ends which allow angular rotation.

Elastic analysis also shows that a column will sustain only small increases in load above the elastic buckling load point. After that, yielding occurs, if it has not already done so, as the column continues to buckle and collapse under the applied load. In essence, buckling theory indicates that there is no significant linear relationship between force and displacement after buckling occurs.

In fact, the Euler buckling load is simply a bifurcation point as discussed in chaos theory. Below the Euler buckling load, the relationship between load and displacement is linear and deterministic. At the Euler buckling load and slightly above it, the column can assume several deflection modalities: a half sine wave bowed outward, a half sine wave bowed inward, an "S" shaped sine wave bowed outward, an "S" shaped sine wave bowed inward, etc. The more complicated modalities are essentially higher frequency sinusoidal functions.

In short columns where the column is a thin-walled metal cylinder or similar, localized buckling is what causes the "accordion-like" folding or crushing of the sheet metal. The buckling occurs where the Euler buckling load is exceeded on a localized basis. As long as the same load which caused the buckling to begin is applied, more column will feed into the "buckling" zone and become crushed. Once started, the process proceeds rapidly and only stops when either the load is removed, or when the column has wholly collapsed into itself.

The above can be easily verified experimentally using thin-walled aluminum soda cans. Most soda cans are nominally about 4-13/16 inches in height, and 2-9/16 inches in diameter. If a load is evenly applied around the top of an undented can and the can is sitting on a firm, flat surface, the can will compress and rebound elastically as long as the load is less than 118 pounds. When the load is at or slightly exceeds 118 pounds, the can will collapse to a height of about 3 inches.

Because the can has reinforced ends, at about a height of 3 inches the end conditions at the top and bottom of the can begin to exert influence. At that height it is necessary to increase the applied load to cause further collapse. As noted before, a column with fully fixed ends has a Euler buckling load twice that of one which has fully pinned ends. A generalized graph of what occurs is shown in Figure 5.

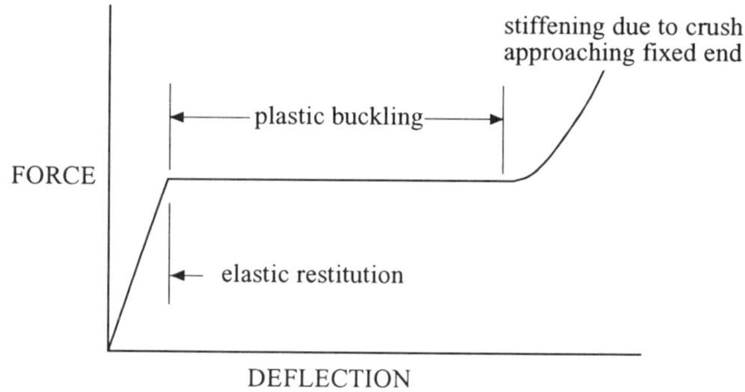

Figure 5. Generalized Graph of Buckling in Short, Thin Walled Column

However, even with the change in end conditions as noted, an aluminum soda can will collapse nearly 40% of its height in response to an applied constant force. Thus, the amount of work necessary to cause the can to crush no more than 40% of its original height is simply 118 pounds times the amount of crush. Since energy units are more often expressed in lbf-ft than lbf-in, then this relationship can be restated as follows:

$$U = k\,c \tag{1}$$

where

U = amount of irreversible work done in crushing the can,
c = the amount of crush in inches, and
k = the force needed for buckling, which in this case is 118 lbf, or 9.833 lbf-ft/in.

Usually, the "k" term is simply called the stiffness coefficient.

H. Going from Soda Cans to the Old Can You Drive

Many people jokingly refer to their cars as an "old tin can." In principle, however, an empty soda can does have similarities to a vehicle. Both are thin-walled metal shells. Both exhibit buckling failure in response to loading before they experience traditional compressive stress material rupture (i.e., they crush before they break). Both have crush load thresholds that depend on the point of application. The force needed to crush a can by pushing down on its top is different from the force needed to crush it by pushing on its side.

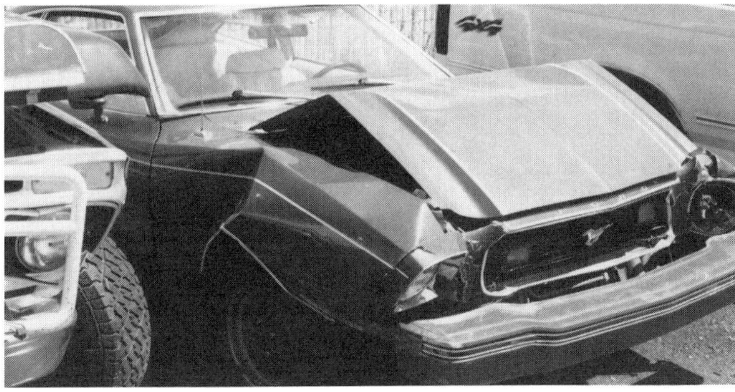

Figure 6. Typical front end buckling due to impact.

If the energy required for front-end crush in a vehicle follows the general form shown in equation (l), then the amount of crush expected in a fixed barrier impact versus impact speed is derived as follows:

$$U = k\,c \qquad \text{(li)}$$

$$\tfrac{1}{2}\, m v_{eq}^2 = k\,c$$

$$v_{eq} = [2kc/m]^{1/2} = [2k/m]^{1/2}\,[c]^{1/2}$$

where v_{eq} = impact speed with a fixed barrier.

If it is assumed that crush begins only above a certain impact speed, where the vehicle's kinetic energy exceeds the elastic energy portion of its force-deflection function, then the above can be modified to the following:

$$v_{eq} = v_o + [2k/m]^{1/2}\,[c]^{1/2} \qquad \text{(lii)}$$

where

v_{eq} = impact velocity with a fixed barrier,
v_o = impact velocity above which damage occurs,
c = average crush depth across front of car, and
2k/m = essentially a constant associated with the properties of the vehicle.

In equation (lii), if "$c = 0$," then "$v_{eq} \le v_o$."

Equations (li) and (lii) above contain a square root of "c," the crush depth. Because of this, the graph of "v_{eq}" versus "c" follows the general plot of a square root function like $y = [ax]^{1/2}$. Thus, the graph of "v_{eq}" versus "c" would look linear at low values of "c," but would then flatten out somewhat at high impact velocities.

In more mathematical terms, the slope of the function "$v_{eq} = f(c)$" is greater at low values of "c" than high values. If the derivative of equation (lii) is taken, the reason for this becomes apparent.

$$dv_{eq}/dc = [k/2m]^{1/2}\,[1/c]^{1/2} \qquad \text{(liii)}$$

As "c" decreases below the value "$[k/2m]^{1/2}$" the slope increases. As "c" increases more than the value "$[k/2m]^{1/2}$", the slope decreases. In a vehicle with a "k" value of about 6,000 lbf-ft/in or 500 lbf, and a mass of 106 lbf-sec²/ft, the term "[k/2m]" computes to a value of 1.54 ft or 18.4 inches. Thus, in this example, the slope of "dv_{eq}/dc" is 45 degrees when crush is 18.4 inches deep.

Equations (li) and (lii) are also interesting concerning the accuracy of calculating the fixed barrier impact velocity from the average crush depth. If the crush were 6 inches, the corresponding impact velocity computes to 26.06 ft/sec. A 1% increase in crush causes the computed speed to be 26.19 ft/sec, a 0.5% increase. Similarly if the crush were 30 inches deep, the corresponding velocity computes to 58.28 ft/sec. A 1% increase in crush causes the computed speed to be 58.57 ft/sec, a 0.5% increase. In short, the square root of "c" tends to halve the error of calculating "v_{eq}" from "c."

This is because of the following relation:

$$[1 + \Delta x]^{1/2} \sim 1 + (\Delta x/2)$$

where $\Delta x >> 1.0$.

In equation (lii), it should be noted that the "v_o" term is essentially the velocity associated with overcoming the elastic energy portion of the vehicle impact. This should be used only if the elastic energy of rebound is not accounted for elsewhere. When the elastic rebound is accounted for, such as calculating the energy expended in the rebound skid, the "v_o" term should be left out to avoid "double bookkeeping."

In general, because the elastic energy component is very small with respect to moderate or high speed collisions, no significant accuracy is lost if the elastic energy term is simply ignored. As noted in Example Problem 2 of Chapter 9, a 1979 Ford Mustang impacting a fixed barrier will expend about 99% of its kinetic energy in crush. Thus, if 1% of the total energy is ignored, the accuracy of the impact speed calculation will be affected as follows:

$$U = \tfrac{1}{2} mv^2 \tag{liv}$$

$$v = [2U/m]^{1/2}$$

$$(v_{approx})/(v_{actual}) = [0.99/1.00]^{1/2} = 0.995.$$

As can be seen by inspection and considering some of the other variables in these calculations, this is a very reasonable simplification for moderate or high-speed impacts.

I. Evaluation of Actual Crash Data

In some crash tests, a 1980 Chevrolet Citation was run into a fixed barrier at 35, 40, and 48 mph. The residual crush across the front was measured to be 21.4, 27.9, and 40.4 inches respectively. The weight of the Citation was 3,130 pounds in each case. From this information, the following table was made.

Table 3. 1980 Citation Crash Data•

impact speed, mph	kinetic energy, lb-ft	crush "c", in.	calc. stiffness "k", lb-ft/in
35	128,200	21.4	5,990
40	167,400	27.9	6,001
48	241,100	40.4	5,968

average "k" = 5,986 lb-ft/in
max. deviation from average = 0.3%

(•Note: Data from Field Accidents: Data Collection, Analysis, Methodologies and Crash Injury Reconstruction, 1985, paper 850437, "Barrier Equivalent Velocity, Delta-V and CRASH3 Stiffness in Automobile Collisions" by Hight, Hight, and Lent-Koop, figure 4.)

The data in Table 3 agree very well with the theory over a range of moderate speeds.

In similar fixed barrier crash tests a 1977 Honda Civic was frontally impacted at 19.1, and 26.8 mph. The weight of the Civic was 1,641 pounds. The residual crush across the front was measured to be 9.5, and 14.3 inches respectively. Table 4 depicts the data.

Table 4. 1977 Civic Crash Data•

impact speed, mph	kinetic energy, lb-ft	crush "c", in.	calc.stiffness "k", lb-ft/in
19.1	20,011	9.5	2,106
26.8	39,397	14.3	2,755

average "k" - 2,430 lb-ft/in
max. deviation from average = 13%

(•Note: Data from "Elastic Properties of Selected Vehicles," by Navin, MacNabb, and Miyasaki, SAE paper 880223.)

The data in Table 4 generally agree with the theory. Due to the form of the equation for velocity, the 13% variation from the average "k" translates to a variation of about 6–7% in the calculation of the impact speed.

In two tests of a 1975 Honda Civic, it was found that the first one crushed 10.0 inches in a 19.7 mph frontal impact with a fixed barrier. In the second, a crush of 12.5 inches was reported for a 23.0 mph impact. The vehicle weighed 1,535 pounds. Table 5 depicts the data.

Table 5. 1975 Civic Crash Data•

impact speed, mph	kinetic energy, lb-ft	crush "c", in.	calc.stiffness "k", lb-ft/in
19.7	19,916	10.0	1,992
23.0	27,148	12.5	2,172

average "k" = 2,082 lb-ft/in
max. deviation from average = 4.3%

(•Note: Data from "Elastic Properties of Selected Vehicles," by Navin, MacNabb, and Miyasaki, SAE paper 880223.)

Table 5 also has excellent agreement with the theory, although the range of velocities involved are not very far apart.

J. Low Velocity Impacts - Accounting for the Elastic Component

As noted in the previous section, when the impact velocity is well above the point where plastic deformation occurs, that is crush, the elastic energy component can be ignored if it is accounted for in the rebound skid or similar. The idea is that any residual elastic energy will eventually be dissipated by other means. However, if it cannot be accounted for in that way, and it is necessary to obtain the impact speed from only the impact crush, then when the elastic energy constitutes perhaps 15% of the total impact kinetic energy, some allowance for it should be made.

For example, consider a car that impacts against a fixed barrier and only has 1 inch of deformation across the front. If the stiffness coefficient "k" is 5,000 lb-ft/in, and the vehicle weighs 2,600 pounds, then it might be estimated that the impact speed was as follows:

$$v_{eq} = [2(5{,}000 \text{ lb-ft/in})(1 \text{ in})/(80.8 \text{ lb-sec}^2/\text{ft})]^{1/2} = 11 \text{ ft/sec}$$

However, since the impact speed is close to the elastic-plastic impact boundary, a more accurate estimate of the impact speed should include the elastic energy expended in rebound effects that are otherwise unaccounted for. Since it is known that this model car will not show crush at fixed barrier impacts of less than 5 mph, then the elastic energy component is:

$$E_{elas} = \tfrac{1}{2} m v_o^2 = 2{,}172 \text{ lb-ft}$$

Combining this with the previous expression for "v_{eq}" gives the following:

$$v_{eq} = [2(5{,}000 \text{ lb-ft} + 2{,}172 \text{ lb-ft})/(80.8 \text{ lb-sec}^2/\text{ft})]^{1/2} = 13.3 \text{ ft/sec.}$$

Thus, allowing for the elastic component in this case increases the calculated impact velocity by about 21%.

K. Representative Stiffness Coefficients

When automobiles are arranged in categories of weight, it is found that the stiffness coefficients of the various vehicles are similar.

The following tables show some of these categories and their representative stiffness coefficients. The data was prepared from crash data, where impacts to a fixed barrier occurred at about 35 mph.•

(•Note: Data from Field Accidents: Data Collection, Analysis, Methodologies and Crash Injury Reconstruction, 1985, paper 850437, "Barrier Equivalent Velocity, Delta-V and CRASH3 Stiffness in Automobile Collisions" by Hight, Hight, and Lent-Koop, figure 4.)

Table 6. Small Car Stiffness Coefficients

vehicle	weight, lbs	stiffness coeff., lb-ft/in
1979 Honda Civic	2,180	4,720
1979 Ford Fiesta	2,190	4,040
1979 Plymouth Champ	2,310	4,260
1979 Datsun 210	2,430	3,960
1979 VW Rabbit	2,600	4,860
1979 Toyota Corolla	2,650	5,340
1979 Chevette	2,730	4,619
AVERAGE	2,441	4,620
RANGE	+12/-11%	+16/-14%

Table 7. Medium Car Stiffness Coefficients

vehicle	weight, lbs.	stiffness coeff., lb-ft/in
1979 Mustang	3,070	7,610
1979 Mercury Capri	3,070	7,178
1979 Chevrolet Monza	3,240	5,970
1979 Volvo 242	3,290	4,600
1979 Ford Fairmont	3,300	6,000
1979 Volvo DL	3,350	5,040
1979 Volvo 244 DL	3.370	4,960
AVERAGE	3,241	5,908
RANGE	+4/-5%	+28/-22%

Table 8. Full Sized Car Stiffness Coefficients

vehicle	weight, lbs	stiffness coeff.,lb-ft/in
1980 AMC Concord	3,700	7,460
1979 Plymouth Volare	3,820	7,170
1979 Olds Cutlass	3,820	5,600
1979 BMW 528	3,840	6,400
1979 Ford Granada	3,950	6,145
1979 Mercury Marquis	4,220	6,300
1979 Ford LTD	4,370	6,850
1979 Dodge St.Regis	4,460	6,470
1979 Olds Regency	4,710	7,355
1979 Ford LTD II	4,810	6,000
1979 Lincoln Cont.	5,360	7,384
AVERAGE	4,278	6,649
RANGE	+25/-14%	+12/-16%

As the tables show, the stiffness coefficient generally increases with the curb weight of the vehicle. This makes sense, since heavier vehicles usually contain more sheet metal making them structurally stronger. In fact, a rough estimate of the stiffness coefficient can be made by using the following empirical equation:

$$k = W(1.75 \text{ ft/in}) \quad \text{(lv)}$$

where

k = frontal stiffness coefficient, and
W = curb weight of the vehicle in pounds.

The side stiffness coefficient for a car ranges from 9,000 lb-ft/in for small cars to 11,500 lb-ft/in for large cars.

The rear stiffness coefficient for a car ranges from 3,500 lb-ft/in for older cars to 6,000 lb-ft/in for late model cars.

When a motorcycle impacts the side of a car, the stiffness coefficient for the impact depth is reportedly about 1,200 lb-ft/in for a 200-pound motorcycle. For a 350-pound motorcycle, the stiffness coefficient is reportedly about 1,750 lb-ft/in.

Figure 7. Frontal buckling. Impact was on the right front of car

L. Some Additional Comments

As noted earlier, the stiffness coefficient is basically the force at which buckling occurs. For example, a car with a frontal stiffness coefficient of 6,000 lb-ft/in has a Euler buckling load of 500 lb. However, it should be recognized that this is the force that has to be applied evenly across the front of the car. Thus, when the impact across the front is uneven, an average depth is determined by making depth measurements at regular intervals and averaging them, as shown in Figure 8.

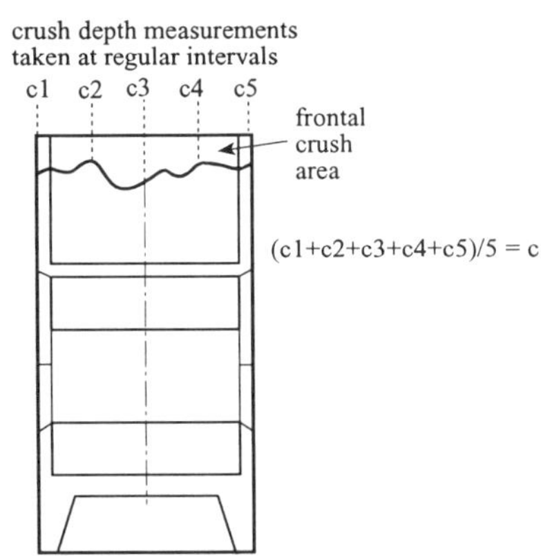

Figure 8. Uneven Frontal Crush

With regard to side crush, this is usually taken to mean the average crush over the area in which impact occurred, especially the area of the vehicle located between the front and rear wheel wells. This is often simply the width across the front of the impacting vehicle.

Figure 9. Side Impact

Because the elastic energy, whether it be a front, side or rear impact, is usually only a very small component of the total energy consumed in an impact, a reasonable approximation in many instances is to assume that the energy expended in the crush of one vehicle is the same as that expended in the other.

Chapter 10: Example Problems

Problem 1

Vehicle "A" is parked. Vehicle "B" runs into the rear end of vehicle "A." Vehicle "B" is equipped with a bumper that is rated to withstand a 5-mph impact with a fixed barrier without sustaining damage. Vehicle "A" is equipped with a bumper that is rated to withstand a 3-mph impact with a fixed barrier without sustaining damage. Vehicle "B" weighs 2,200 pounds, and vehicle "A" weighs 2,800 pounds. What is the maximum speed that vehicle "B" could have had at impact with vehicle "A?"

Solution

The set up equation is as follows:

$$\tfrac{1}{2}\, m_B v_B^{\,2} = E_{A,\ elastic} + E_{B,\ elastic}$$

The elastic energy absorbed by the bumpers at the barrier equivalent impact speed ratings are as follows:

$$E_{A,\ elastic} = \tfrac{1}{2}\,(87 \text{ lbf-sec}^2/\text{ft})(4.4 \text{ ft/sec})^2 = 842 \text{ lbf-ft}$$

$$E_{B,\ elastic} = \tfrac{1}{2}\,(68.4 \text{ lbf-sec}^2/\text{ft})(7.33 \text{ ft/sec})^2 = 1{,}838 \text{ lbf-ft.}$$

Thus, the solution is then:

v_B^2 = (2)(842 lbf-ft + 1,838 lbf-ft)/(68.4 lbf-sec^2/ft) = 78.36 ft^2/sec^2

v_B = 8.85 ft/sec = 6 mph.

Even though the rating of the bumper of vehicle "B" was 5 mph, the vehicle could withstand a 6-mph impact with vehicle "A." This is because vehicle "A" is not a fixed barrier; it also has a bumper that absorbs energy up to a point without sustaining damage. Thus, the kinetic energy of vehicle "B" could be absorbed by both vehicles.

Problem 2

A car weighs 2,500 pounds and has a front-end stiffness coefficient of 6,000 lbf-ft/in. It skids 50 ft and then crashes into a rigid stone retaining wall. When impact occurred, the car was at a slight angle to the wall so that the impact damage at the front right corner is 6 inches of crush, and at the front left corner is 10 inches of crush. Between the two corners, the crush boundary follows a line between the two corners. At what speed was the car going just prior to the initiation of skid?

Solution

The energy method generalized solution to this problem is as follows:

$\frac{1}{2} mv^2 = (mg)fd + kc_{ave}$

where

m = mass of car,
v = velocity prior to skid,
f = coefficient of friction with the road,
d = skid length,
k = stiffness coefficient, and
c_{ave} = average front end crush depth.

The solution to the above equation is:

$v^2 = (2)gfd + (2)kc_{ave}/m$

v^2 = (2)(32.17 ft/sec^2)(0.7)(50 ft) + (2)(6,000 lb-ft/in)(8 in)/(77.7 lbf-sec^2/ft)

v = 59 ft/sec = 40 mph.

Problem 3

Two vehicles collide head on and stick together. The force of the collision is such that the tires on both vehicles go flat. If the crush on each car is 22 inches deep, each car weighs 2,800 pounds, the height of the center of gravity is 28 inches, the axle to axle length is 92 inches, and the stiffness coefficient of each car is 5,500 lbf-ft/in, what is the downward force which causes the tires to go flat?

Solution

Since the collision is head on, and the amount of crush in both vehicles is identical, the collision is identical to a fixed barrier collision of a single vehicle.

The impact velocity is calculated as follows:

$$\tfrac{1}{2}\,mv^2 = (5{,}500 \text{ lbf-ft/in})(22 \text{ in}) = 121{,}000 \text{ lbf-ft}$$

$$v^2 = (2)(121{,}000 \text{ lbf-ft})/(87 \text{ lbf-sec}^2/\text{ft}) = 2{,}782 \text{ ft}^2/\text{sec}^2$$

$$v = 53 \text{ ft/sec} = 36 \text{ mph}$$

Since the vehicle decelerated from 53 ft/sec to zero in 22 inches, then the rate of linear deceleration is calculated as follows:

$$a = v^2/(2d)$$

$$a = (53 \text{ ft/sec})^2/(2)(1.833 \text{ ft}) = 766 \text{ ft/sec}^2 = 23.8 \text{ g's.}$$

Assuming the above, then the time interval for the deceleration is as follows:

$$t = [2d/a]^{1/2}$$

$$t = [2(1.833 \text{ ft})/(766 \text{ ft/sec}^2)]^{1/2} = 0.069 \text{ sec.}$$

The change in momentum is:

$$\Delta M = (53 \text{ ft/sec})(87 \text{ lbf-sec}^2/\text{ft}) = 4{,}611 \text{ lbf-sec}$$

The impulse is then:

$$I = F(t) = \Delta M = 4{,}611 \text{ lbf-sec}$$

Since the time interval is 0.069 seconds, then the force (assuming it is constant over the interval) is as follows:

$F(t) = 4{,}611$ lbf-sec

$F = (4{,}611 \text{ lbf-sec})/(0.069 \text{ sec}) = 66{,}826$ lbf

If the torque externally applied by the frontal impact is just balanced by the torque which tries to push the front end down, then the force pushing the front end down can be calculated.

$T_i = T_j$

$(66{,}826 \text{ lbf})(1.833 \text{ ft}) = F(7.67 \text{ ft}) + (2{,}800 \text{ lbf})(0.60)(7.67 \text{ ft})$
where rotation is about the rear wheels.

$F = 14{,}290$ lbf

Since there are two wheels, then each wheel carries about 7,145 pounds in addition to the usual weight. Since the front axle carries 60% of the weight of the car, then the usual load for each front wheel is:

$(2{,}800 \text{ lbf})(0.60)/2 = 840$ lbf.

Thus, the downward force caused by the head-on collision momentarily loads each front wheel about 8.5 times more than its normal load.

This is the reason why in severe head-on collisions it is not unusual for the front tires to go flat, even if they are not directly impacted.

Problem 4

The wood in a particular type of telephone pole has a shear strength of about 3,000 lbf/in^2. The telephone pole has a diameter of 12 inches. A loaded pickup truck with a total weight of 9,500 pounds skids 22 feet on dry asphalt, collides with the telephone pole and shears it off. Essentially, after impacting the telephone pole, the truck comes to a stop. Inspection of the truck front afterwards finds that an extremely well-reinforced bumper made of H-beams had been welded onto the front in order to push cars and earth-moving machinery. This bumper exhibited no significant crush damage.

How fast was the truck going when it began skidding? How fast was it going when it collided with the telephone pole? What was the g-force on the driver, who was properly wearing a seat belt and harness?

Solution

In this problem there are two main types of irreversible work: the skid, and the shearing of the telephone pole.

U_{skid} = (9,500 lbf)(0.65)(22 ft) = 135,850 lbf-ft.

U_{shear} = (3,000 lbf/in^2)(12 in)^2(3.14159/4)(12 in) = 339,292 lbf-ft

The initial velocity is then:

$\frac{1}{2} mv_1^2 = U_{skid} + U_{shear}$

v_1 = [2(475,142 lb-ft)/(295 lbf-sec^2/ft)]$^{1/2}$ = 56 ft/sec = 39 mph.

Since the skid dissipated 135,850 lbf-ft of energy, then the velocity at impact with the telephone pole is as follows:

$\frac{1}{2} m[v_1^2 - v_2^2]$ = 135,850 lbf-ft

$[v_1^2 - v_2^2]$ = 2(135,850 lbf-ft)/(295 lbf-sec^2/ft) = 920 ft^2/sec^2

v_2 = 47 ft/sec = 32 mph.

The final deceleration was from 47 ft/sec to zero in 12 inches.

$a = v_2^2/2d$

a = (47 ft/sec)^2/(2)(1 ft) = 1,105 ft/sec^2 = 34.3 g's.

This is an extremely high deceleration even for a person who is wearing a seat belt and harness. It is likely that severe, and possibly fatal, injuries would occur in such an impact. The g-force decelerations that the driver would experience can be compared to representative values listed in the head injury criteria (HIC) or upper body severity index (SI). Such comparisons indicate probable types and severity of injuries.

Problem 5

Car "A" travels due north through an intersection. Car "B" travels due east through the same intersection. Car "B" impacts car "A" in its left side, more or less in the area of the driver's door. All of the damage to car "B" is to its front.

Just prior to impact, the driver of car "A" saw the impending collision and tried to stop. Thus, car "A" left a 12-foot skid mark prior to impact directed due north, and a 10-foot skidmark directed in a northeasterly direction after impact. The left side crush in car "A" was measured to be an average of 5 inches deep over an area about 76 inches long.

Just prior to impact, the driver of car "B" also attempted to stop and left a 5-foot skid mark. The front end of car "B" was crushed to an average depth of 9 inches. After impact, car "B" skidded another 10 feet more or less with car "A".

What were the speeds of the two vehicles prior to impact?

Solution

In this problem, it is important to recall that the work done is always in the direction of travel. Work is not done perpendicular to the direction of travel. This is because:

$$U = F \cdot d \quad \text{or} \quad |U| = |Fd \cos \alpha| \qquad \text{vector dot product}$$

Thus, in this case, the crush in the left side of car "A" is attributed to the kinetic energy of car "B." Car "A" did not help cause this crush work.

However, there was work done between cars "A" and "B" due to the metal-to-metal friction of car "A" moving north as car "B" was in contact with the side. In this case, half of the work is attributed to car "A" and half to car "B."

Assumptions:

coefficient of friction on pavement of each car	$f_p = 0.7$
coefficient of friction of metal to metal	$f_m = 0.3$
side stiffness coefficient	$k_A = 11{,}000$ lbf-ft/in
frontal stiffness coefficient	$k_B = 6{,}000$ lbf-ft/in
weight of car "A"	$W_A = 3{,}200$ lbf
weight of car "B"	$W_B = 2{,}800$ lbf
width of car "B"	$l = 64$ in

The energy equation for car "A" is then as follows:

$E_A = \frac{1}{2} m_A v_{A1}^2 =$
$W_A(12 \text{ ft} + 10 \text{ ft})(0.7) + \frac{1}{2}(76 \text{ in} - 64 \text{ in})(1 \text{ ft}/12 \text{ in})(F)$

In the last term, the force during contact "F" is taken to be the force needed for buckling, which is the stiffness coefficient. Thus the energy equation for vehicle "A" is:

$E_A = \frac{1}{2} m_A v_{A1}^2 =$
$W_A(12 \text{ ft} + 10 \text{ ft})(0.7) + \frac{1}{2}(76 \text{ in} - 64 \text{ in})(1 \text{ ft}/12 \text{ in})(917 \text{ lbf})$

Solving for the velocity just prior to when skidding began gives the following:

$E_A = 49{,}738$ lbf-ft

$v_{A1} = 31.6$ ft/sec $= 21.6$ mph

The energy equation for vehicle "B" is set up similarly except that the two crush terms are added. The two crush terms are attributed to vehicle "B" only because vehicle "A" had no crush caused by it in its direction of travel.

$E_B = \frac{1}{2} m_B v_{B1}^2 =$

$W_B(5 \text{ ft} + 10 \text{ ft})(0.7) + \frac{1}{2}(76 \text{ in} - 64 \text{ in})(917 \text{ lbf})$
$+ (9 \text{ in})(6{,}000 \text{ lbf-ft/in}) + (5 \text{ in})(11{,}000 \text{ lbf-ft/in})$

Solving the above gives the following:

$E_B = 138{,}858$ lbf-ft

$v_{B1} = 56.5$ ft/sec $= 38.5$ mph

The velocities at impact can easily be obtained by dropping out the skids on both vehicles which occurred prior to impact. In vehicle "A" this was 12 ft and in vehicle "B" this was 5 ft. It is left to the reader to do this.

Knowing the angle at which both vehicles traveled after impact would also allow this problem to be solved with momentum equations and coefficient of restitution methods. This would be an excellent cross check.

One other point is worth noting. The friction work done by the two vehicles abrading one another during contact was calculated in this example as noted. The work attributed to car "A" was only about 0.9% of its total kinetic energy prior to skidding, and only 0.3% of the total kinetic energy of car "B" prior to skidding. In most cases, the omission of this small amount of work does not introduce any significant error.

CHAPTER 11

Curves and Turns

A. Transverse Sliding on a Curve

When a vehicle executes a turn, the vehicle experiences a centripetal force in accordance to the following:

$$F_c = mv^2/R = m\omega^2R \quad \text{(i)}$$

where

F_c = centripetal force due to rotational motion,
R = turning radius,
v = tangential velocity of vehicle, i.e., forward speed of vehicle,
m = mass of the vehicle, and
ω = angular velocity

When the centripetal force on the vehicle is significant, it can cause the vehicle either to slip off the roadway, or to turn over. If the centripetal force is less than what is required to cause the vehicle to turn over, but more than the sideways frictional force of the tires, the vehicle will slide. Ignoring turnover for the moment, a slide will occur when:

$$F_c \geq F_f \quad \text{(ii)}$$

where

F_c = centrifugal force, and
F_f = sideways frictional force.

Appropriate substitution into equation (ii) above gives the following:

$$mv^2/R \geq Wf \qquad \text{(iii)}$$

$$v^2/(gR) \geq f \text{ or } v \geq [gRf]^{1/2}$$

where
f = coefficient of friction,
g = gravitational constant, and
W = weight of vehicle.

The meaning of equation (iii) is simple: if the value of "$v^2/(gR)$" is greater than the value of "f", the vehicle will lose traction and slide. If this occurs, the direction of slide will be tangent to the curve at the point where the centripetal force exceeded the restraining sideways frictional force.

Equation (iii) also explains why it is best to steer out of a curve rather than to steer into it. During a turn, steering into the turn causes the value for "R," the radius of curvature, to decrease. A decrease in "R" in equation (iii) causes the value of "mv^2/R" to increase. If the vehicle was already near the point of losing traction, a small decrease in "R" may be sufficient to increase the centripetal force to the point where it exceeds the frictional forces holding it. However, steering out of the curve causes the value of "R" to increase, which decreases the centripetal force.

Also, if a vehicle is taking a turn near the point at which sliding may occur, any lessening of the side coefficient of friction can cause the vehicle to slide out of the turn. Such reductions in the side coefficient of friction can not only be caused by variations in the pavement, but also by braking. As noted in Chapter 4, when there is significant slip in the forward or longitudinal direction, the coefficient of friction in the transverse direction decreases. Thus, braking during a tight turn can cause a vehicle to lose side traction and slide out of the turn.

This is the reason why it is sometimes recommended that a driver accelerate slightly during a turn and not attempt any significant braking. By accelerating slightly through a turn, the amount of slip between the tires and the pavement is minimized, and the lateral coefficient of friction remains as high as possible.

Professional drivers, especially on dirt tracks, often use these principles in concert to turn corners faster than otherwise. For example, when approaching a turn, the driver will cause the rear wheels to lose traction and the rear end of the vehicle will swing around. When the rear end has swung around to the point where the car is pointed towards the inner radius of the turn, the driver will steer out of the slide and then accelerate. In a sense, instead of steering through the turn, the driver literally rotates his vehicle about the front wheels. This allows a turn to be taken without slowing down. (Like all other such maneuvers discussed in this book, it is not recommended that this be done by an amateur.)

So far, however, it has been assumed that the roadway is flat. In many cases, a curved roadway will be banked, that is, set so that the roadway has a certain pitch or angle in the transverse direction. If the roadway has been well designed, it will be banked so that if the curve is driven at the recommended speed, the centripetal force effects will just be balanced out by the angle of the bank, or the superelevation as it is sometimes called.

If the angle of the bank just cancels out the centripetal force, the following holds:

$$(mv^2/R)(\cos\theta) = W(\sin\theta) \qquad \text{(iv)}$$

where θ = angle of bank of the roadway.

Solving equation (iv) for "q" we find that:

$$\tan\theta = (v^2)/(gR) \qquad \text{(v)}$$

$$\theta = \arctan[(v^2)/(gR)]$$

If the angle is small, say less than about 8 degrees, then the small angle approximation can be applied where

$\tan\theta \sim \theta$ when the angle is measured in radians.

Of course, since the definition of measurement in radians is the amount of circumference of a circle divided by its radius, then

$$\theta = h/l$$

where

h = the height of the roadway above level, and

l = the length across the roadway.

Thus, when the angle of bank is 8 degrees or less, the following holds:

$$h/l = \theta = (v^2)/(gRf) \qquad \text{(vi)}$$

If it is assumed that the roadway is banked positively (the angle of bank helps keep the vehicle from side sliding during the curve), then the general expression for slippage is as follows:

$$(mv^2/R)(\cos\theta) \geq Wf(\cos\theta) + W(\sin\theta) + (mv^2/R)(\sin\theta)f$$

$$v^2 \geq [gRf + gR(\tan\theta]/[1 - f(\tan\theta)]$$

$$v_2 \geq [gRf + gR(h/l)]/[1 - f(h/l)] \qquad \text{(vii)}$$

Since the term "f(h/l)" is small and becomes even smaller under the square root radical, the above expression for turns on banks of 8 degrees or less can be simplified to the following:

$$v \geq [gR(f + h/l)]^{1/2} \qquad \text{(viii)}$$

When the banking angle is 8 degrees, or "h/l = 0.14," the error in the above equation is about -5%, that is, equation (viii) will compute a speed that is about 5% too low when the angle of bank is 8 degrees. Since most roadways are banked at angles much less than 8 degrees, the approximation is reasonably accurate for most purposes.

However, sometimes an older roadway or a very poorly designed one will have negative banking. Negative banking is the case where the angle of the bank reduces the speed at which a turn can be made. When there is negative banking, equation (vii) has a change in sign in the second and third terms on the right hand side of the inequality as follows:

$$(mv^2/R)(\cos\theta) \geq Wf(\cos\theta) - W(\sin\theta) - (mv^2/R)(\sin\theta)f \qquad \text{(ix)}$$

Equation (ix) can be similarly reduced to the following:

$$v_2 \geq [gRf - gR(h/l)]/[1 + f(h/l)] \text{ or} \qquad \text{(x)}$$

$$v \geq [gR(f - h/l)]^{1/2} \text{ when } f(h/l) \text{ is small.}$$

Equation (x) shows that if the coefficient of friction "f" becomes less than "h/l," the vehicle can slide down the bank without having any forward motion at all.

The above relations can be useful in setting a lower-bound limit to the speed of a vehicle if it slid out of the curve. If there is a question as to whether the vehicle slid out of the curve due to excessive speed for the curve, or just by poor operation of the brakes at a lower speed, the skid marks can be examined. Skid marks which are initiated by a braking action will generally have longitudinal striations left by the tire treads. Skid marks left by lateral slide caused by excessive speed in rounding the curve will usually have barber pole appearing striations, showing that the tire was still rotating forward as the transverse sliding occurred. Such marks are usually call yaw marks or yaw skid marks.

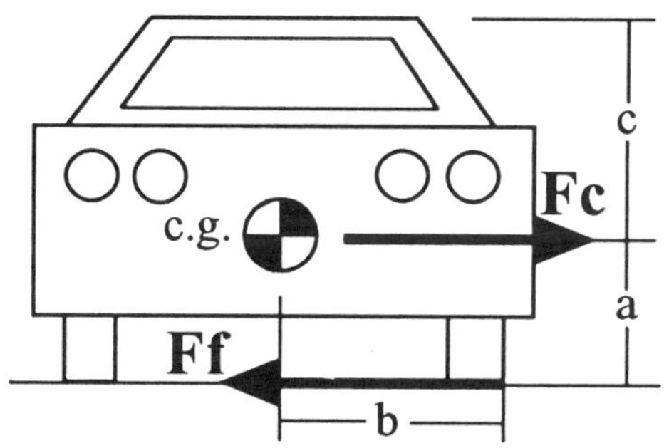

Figure 1. Forces During a Turn

B. Turnovers

A turnover occurs during a turn or curve when the centripetal force becomes sufficient to overcome the forces which hold the vehicle on the ground. However, in order for a turnover to occur, the vehicle must maintain lateral traction so that lateral slip does not occur first. Referring to figure 1, the moment and force equations that determine when turn over occurs is as follows:

$$(mv^2/R)a > W(b) \text{ and} \tag{xi}$$

$$Wf \geq (mv^2/R)$$

where

W = weight of the vehicle,
m = mass of the vehicle,
g = acceleration due to gravity,
R = radius of curve, and
a and b are dimensions as shown in Figure 1.

Solving equations (xi) gives the following:

$$[gRf]^{1/2} > v > [gR(b/a)]^{1/2} \quad \text{(xii)}$$

If the roadway is positively banked, the conditions for turnover are then modified as discussed in the previous section to the following:

$$[gR(f + h/l)]^{1/2} > v > [gR(b/a)]^{1/2} \quad \text{(xiii)}$$

Similarly, if the roadway is negatively banked, the conditions for turn over are as follows:

$$[gR(f - h/l)]^{1/2} > v > [gR(b/a)]^{1/2} \quad \text{(xiv)}$$

Equation (xiii) contains the basic information necessary to determine whether a vehicle will turn over first, or slide first. Rearranging equation (xiii) gives the following relation which defines the conditions under which turn overs occur:

$$(f + h/l) > v^2gR > (b/a) \quad \text{(xv)}$$

or $(f + h/l) > (b/a)$

A typical compact car will have an "a" value of about 2 feet, and a "b" value of about 2.58 ft. Thus, in order for such a car to turn over while making a curve, the coefficient of friction must be greater than 1.29. Since a typical coefficient of friction for a dry, very soft tire is no more than about 0.80, this means that under most circumstances, a typical compact car will slide out of a turn rather than turn over. Of course, a coefficient of friction of 1.29 or greater can be accomplished by side impact with a curb or other type of low barrier. In such cases, lateral sliding is simply prevented by the barrier.

The maximum coefficient of friction for a soft truck tire on dry concrete can be as high as 0.94. This means that a truck or trailer with a "b/a" ratio of 0.94 or less could turn over in a turn. Since a common width of a truck or trailer is about 80 inches, that means that it is possible for a truck or trailer whose center of gravity is 43 inches or more off the ground to turn over rather than slide. Since the floor height of commercial trailers is often four feet or more and cargo is stacked on top of the floor, commercial trailers can turn over rather than slide in a turn. This is especially true if the trailer is stacked to the ceiling with cargo and is carrying the maximum load. This greatly shifts the center of gravity upwards.

Further, if the load in a trailer is not centered over the wheels, then the center of gravity will not be located halfway between the wheels as is assumed. If a heavy load, for example, were stacked along one side of the trailer, the "b" value might be much shorter than half the width of the trailer. This would further increase the tendency of the trailer to tip over during a turn.

C. Load Shifting

It often occurs that when a trailer turns over on a curve, the driver blames "load shift" rather than excessive speed. "Load shift" happens when the load being carried shifts position during the turn, causing the center of gravity to shift to an unfavorable position. The shifting load may even impact the wall of the trailer which would further increase the tendency of the trailer to tip. In such cases, it is worth noting that the U.S. Department of Transportation (USDOT) regulations, as contained in the Federal Motor Carrier Safety Regulations, require that the load being carried be positioned in the most favorable position to minimize the center of gravity (section 393.100). The regulations also require that the load be secured to prevent "load shifting."

Because of the common use of "load shifting" to explain trailer turn overs on curves, the following USDOT regulation (section 392.9) is quoted in whole:

> "(a) General. No person shall drive a motor vehicle and a motor carrier shall not require or permit a person to drive a motor vehicle unless ----
> (1) The vehicle's cargo is properly distributed and adequately secured as specified in 393.100–393.106 of this subchapter."

D. Side Versus Longitudinal Friction

The total amount of frictional resistance of a tire has been previously given as:

$$F_f = W\,f \tag{xvi}$$

where

F_f = the friction force resisting motion,
f = coefficient of friction, and
W = the load applied on the tire.

The above is generally true when only one resistive force is at work, like a longitudinal or transverse resistive force. However, when both a side (transverse) and longitudinal resistive force is at work, the total frictional force is the vector sum of the two component frictional forces.

For example, let "F_{long}" be a frictional force being applied in the longitudinal direction, and "F_{side}" be a frictional force being applied along the side or transverse direction. In that case, the following would hold:

$$[F_{side}^2 + F_{long}^2]^{1/2} = F_f = Wf, \text{ or alternately} \qquad \text{(xvii)}$$

$$[f_{long}^2 + f_{side}^2]^{1/2} = f$$

Thus, if a vehicle is accelerating forward and is using 80% of its available maximum traction to accomplish this, then there would be only 60% of the maximum traction available for side frictional forces.

$$[0.8F_f^2 + F_{side}^2]^{1/2} = Wf = F_f$$

$$F_{side} = 0.60F_f$$

Equation (xvii) is often called the *friction circle*. When longitudinal braking and side friction or traction are being considered simultaneously, however, the friction circle relationship is not exact. Laboratory and field measurements have shown that side friction is usually a little less than what would be predicted strictly by equation (xvii). This is due to tire deformation mechanics. For this reason, the following empirical modification to equation (xvii) is used when braking is involved:

$$[(1.2)f_{side}^2 + f_{long}^2]^{1/2} = f \qquad \text{(xviii)}$$

Using the above relationship, if the tire-pavement coefficient of friction is 0.75, and the vehicle is making a turn such that the side friction just equals the centripetal force, then a braking action causing a 0.2g slow down ($f_{long} = 0.2$) will cause the side coefficient of friction to drop from 0.685 to 0.660, a 3.6% decrease. Since prior to the braking action the side forces just matched the centripetal forces, then the slight decrease in side friction caused by the braking will cause the vehicle to slide out of the turn.

E. Cornering and Sideslip

When a vehicle executes a turn, the front wheels are rotated a certain angle, called the *steering angle*. Unless the vehicle is traveling at low speed, the steering angle and the actual direction of travel of the vehicle will not be exactly the same because of tire slip in both the longitudinal and transverse directions. Usually, the steering angle will be a little greater than the actual angle of turn. The difference between the two angles is called the *angle of slip*.

One way to imagine what is going on is to think about the car going straight forward and then turning. Unless the vehicle is traveling at a very slow speed, the forward momentum of the vehicle will tend to continue to push the vehicle straight ahead, while the front wheels try to turn the vehicle left or right. This results in some slip across the contact "patch" between the tire and the ground in both the longitudinal and transverse directions. Thus, the tire will exert both side and longitudinal forces.

The angle of slip effect can be easily observed by executing the same tight turn at several speeds. At very slow speeds, a car can make a tighter turn at the same steering angle than at higher speeds. In fact, with the steering wheel turned to its maximum left or right position, it can be seen that a car will spiral outward with an increasing radius of curvature as the forward speed is slowly increased.

Most ordinary turns develop slip angles of 5 degrees or less. In this range, the side forces developed by a tire follow the following linear relation:

$$F_{side} = K\lambda \qquad \text{(xix)}$$

where

λ = slip angle, and

K = turning stiffness constant for a specific load, tire type, camber, etc.

Very hard turns at relatively high speeds can have slip angles of 10 to 17 degrees. Unfortunately, above 5 degrees, the relationship between slip angle and side force is not linear. The side force curve tends to flatten out, like a square root curve.

This flattening out effect at high slip angles is sometimes observed at dirt track races, where a driver takes a corner at too sharp a turn. Instead of turning, the car will often just continue forward until it slows down enough to where the tires "bite" the surface again.

In general, radial tires have higher "K" values than bias tires. Increased camber angles of the tires will reduce "K." Tire inflation pressure, and tire load will also increase the value of "K."

For passenger cars, "K" may range from 50 to 200 lbf/deg with an average value being about 100 lbf/deg. For trucks, "K" may range from 200 to 500 lbf/deg. with an average value being about 300 lbf/deg.

F. Turning Resistance

When a vehicle executes a turn, a resistance to forward motion is produced by the slip angle of the tires. If the slip angle is "λ" and the side coefficient of friction is "f_{side}," then the relationship between slip angle, side friction and forward friction is as follows:

$$f = f_{side}\,(W_{front}/W)\sin\lambda \qquad \text{(xx)}$$

where

W_{front} = load carried by the front wheels, and
W = total load of the vehicle carried by both axles.

When the front wheels slip angle is small, the rear wheels have a negligible slip angle, and equation (xx) is appropriate. However, if the turn is large enough so that slip angle in the rear wheels must be taken into account, equation (xx) must be expanded to include the rear wheels as follows:

$$f = f_{side}\,[(W_{front}/W)(\sin\lambda_{front}) + (W_{rear}/W)(\sin\lambda_{rear})] \qquad \text{(xxi)}$$

For example, if a 3,000-lbf vehicle has a cornering stiffness of 100 lbf/deg, the front wheels are turning at a slip angle of 3 degrees and the rear wheels are turning at a slip angle of 1 degree, and the front axle carries 60% of the total load, then the forward turning resistance is as follows:

$$F_{side} = K\lambda = (100\ \text{lbf}/°)(3\ \text{deg}) + (100\ \text{lbf}/°)(1°) = 400\ \text{lbf}$$

$$f_{side} = F_{side}/W = 400\ \text{lbf}/3{,}000\ \text{lbf} = 0.133$$

$$f = (0.133)[(0.60)(\sin 3°) + (0.40)(\sin 1°)] = 0.005$$

In this case, when the total weight of the vehicle is taken into consideration, the total resistive force in the direction of travel caused by turning is:

$$F_f = Wf = (3{,}000 \text{ lbf})(0.005) = 15 \text{ lbs.}$$

This is not particularly significant. In general, at low to moderate slip angles, the effects of cornering with respect to resistance to forward motion is negligible.

G. Turning Radius

The following discussion is limited to two-axle vehicles.

At slow speeds, the turning radius, as measured to the longitudinal axis of the vehicle, is given by the following:

$$R = L/(\varphi_f - \varphi_r) \qquad \text{(xxii)}$$

where

- R = turning radius,
- L = distance between axles (wheel base),
- φ = steering angle (in radians) as measured from longitudinal axis,
- subscript r denotes rear, and
- subscript f denotes front.

In most vehicles, "φ_r" is zero, that is, the rear wheels don't steer. However, there are a limited number of production vehicles that have steerable rear wheels. In such cases, "φ_r" has a value, usually the same as "φ_f."

The angle turned by the steering wheel is usually not the same angle that is turned by the front wheels. The ratio between the steering wheel angle and the front wheels steering angle is called the *steering ratio.*

$$\varphi = \alpha/\varphi \qquad \text{(xxiii)}$$

where

- α = steering wheel angle, and
- φ = steering angle of front wheels.

At speeds where turning slip occurs, the turning radius is given by:

$$R = L/(\varphi_f - \lambda_f - \varphi_r + \lambda_r) \qquad \text{(xxiv)}$$

where

L = wheel base,
φ = steering angle of front wheels (in radians),
λ = slip angle (in radians),
subscript r denotes rear, and subscript f denotes front.

An inspection of equation (xxiv) finds that there are three general possibilities:

1. $\lambda_f - \lambda_r = 0$ *Neutral steering.* The effects of slip angle at the front wheels is cancelled by the slip angle of the rear wheels.

2. $\lambda_f > \lambda_r$ *Under steering.* The turning radius is larger than that indicated by the steering angle.

3. $\lambda_f < \lambda_r$ *Over steering.* The turning radius is smaller than that indicated by the steering angle.

In a vehicle that oversteers, the rear tires will tend to slide before the front tires. In a vehicle that understeers, the front tires tend to slide before the rear tires. In a vehicle with neutral steering, the front and rear tires tend to slide at the same time. Many production vehicles are designed to have a slight oversteer bias.

A vehicle's understeering or oversteering tendencies may be important in making curves near the critical speed for either slip or turnover. For example, in a vehicle that oversteers, if the curve is being taken at an excessive speed, release of the accelerator (and therefore reduction of the slip angle) automatically causes the vehicle to turn at a slightly larger radius without adjustment of the steering angle.

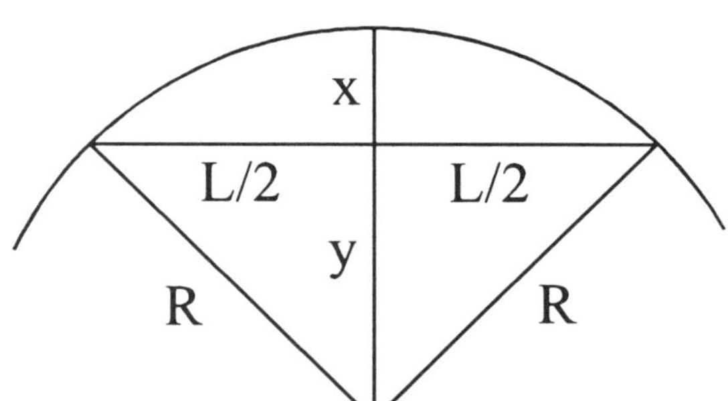

Figure 2. Circle Segment

H. Measuring Roadway Curvature

Referring to figure 2, it can be seen that:

$$y^2 + (L/2)^2 = R^2 \qquad \text{(xxv)}$$

It is also noted that:

$$y = R - x \qquad \text{(xxvi)}$$

Combining equations (xxv) and (xxvi) gives the following:

$$(R - x)^2 + (L/2)^2 = R^2 \qquad \text{(xxvii)}$$

$$R = L^2/(8x) + x/2$$

From a practical standpoint, it is convenient to use a standard length to measure "L," the chord length, and then measure "x," the depth of the chord. If, for example, a 50 ft. tape is used as the standard, equation (xxvii) becomes:

$$R = (2{,}500)/(8x) + x/2 = 312.5/x + x/2 \qquad \text{(xxviii)}$$

where all measurements are in feet.

I. Motorcycle Turns

The height of the center of gravity of a motorcycle and rider is often about 24 inches off the ground. When a motorcycle makes a turn, the lean of the rider and motorcycle must equalize the centripetal force in the following way:

$$(mv^2/R)a(\cos\theta) = W\,a(\sin\theta) \qquad \text{(xxix)}$$

where

θ = angle of lean (as measured from the vertical),
W = weight of the motorcycle and rider,
a = height of the center of gravity when fully upright.

Equation (xxix) reduces to the following:

$$\tan\theta = v^2/gR \qquad \text{(xxx)}$$

$$\theta = \text{Arctan}(v^2/gR)$$

While professional riders may lean as much as 45 degrees from the vertical to take a turn, most riders usually don't lean more than about 20–25 degrees from the vertical, and inexperienced riders may not lean more than 15 degrees from the vertical.

Chapter 11: Example Problems

Problem 1

It is desired to derive a simplified but reasonably accurate equation that gives the critical speed for curves when the bank angle is more than 8 but less than 20 degrees.

Solution

In equation (vii), the substitution of "$\tan\theta \sim \theta$" was made. This is the first term of the general expansion series for "$\tan\theta$." Using the first two terms of the expansion gives the following substitution:

$$\tan\theta \sim \theta + \theta^3/3.$$

It is found that when the angle is 20 degrees or 0.349 radians, the tangent is 0.364. Using the two-term approximation gives an answer of 0.363, which varies only 0.3% from the 0.364 value. Thus, the two term approximation will suffice very well.

Substituting this into equation (vii) gives the following:

$$v^2 \geq [gRf + gR(\tan\theta)]/[1 - f(\tan\theta)]$$

$$v^2 \geq [gRf + gR(h/l + h^3/3l^3)]/[1 - f(h/l + h^3/3l^3)]$$

$$v_2 \geq [gR][3fl^3 + 3hl^2 + h^3]/[3l^3 - 3fhl^2 - fh^3]$$

Since "$l > h$" by about a factor of three, then in relation to "l^3" the term "h^3" is at least 37 times smaller, i.e., it is negligible.

$$v^2 \geq [gR][3fl^3 + 3hl^2] / [3l^3 - 3fhl^2]$$

$$v_2 \geq [gR][f + h/l] / [1 - fh/l]$$

The above is basically identical with equation (vii) already given. When "$f = 0.7$" it is found that using the tangent of 20 degrees gives a value for "v^2" is 2.7% larger than when the above equation is used.

Since the 3% error is under the square root radical, the actual error in calculating "v" will be less than 1.5%.

Problem 2

A particular model car made by Honda has a steering system that has both front wheel and rear wheel steering control. At highway speeds, the front and rear wheels turn at the same angles parallel to each other. At low speeds, the front and rear wheels turn so that they are left-right mirror images of each other.

Discuss the handling characteristics of this vehicle.

Solution

For low speeds where slip is negligible, equation (xxii) applies.

$$R = L/(\varphi_f - \varphi_r)$$

Since the two steering angles are equal but opposite to one another then, equation (xxii) becomes:

$$R = L/(2\varphi)$$

This indicates that for the same steering angle, the vehicle can make a much tighter turn because the turning radius is halved.

For higher speeds where slip becomes significant and the wheels operate parallel to each other rather than mirror image opposite, equation (xxiv) applies.

$$R = L/(\varphi_f - \lambda_f - \varphi_r + \lambda_r)$$

Since the slip angles are the same for both front and back, the slip angles cancel each other out and the steering becomes neutral. Also, because both the front and rear steering angles are the same, then the turning radius is infinite. In other words there is no turning radius as no turn is being made according to the equation.

It should be remembered that equation (xxiv) is based on a static position of the steering wheel, where the wheel is turned to a certain position and then the vehicle moves at a constant rate. In the case of this particular vehicle, when the steady-state steering position is reached, the turn is already over and the vehicle is pursuing a straight-line trajectory. Thus, the turn curvature occurs during the time the steering wheel is turned to the steady state-position, a straight (but slightly diagonal) trajectory is then followed when the steering wheel reaches the steady state position, and then a reverse turn curvature is executed to return the steering wheel to its straight ahead position.

For such a case, the turn curvature is time dependent as follows:

$$\omega R = [v^2 + u^2]^{1/2}$$

where

ω = rate of change per time of steering angle (in radians per second),
v = instantaneous forward speed at time of turn,
u = instantaneous lateral speed at time of turn, and
R = turning radius.

Because the steering angle of the wheels is proportional to the steering angle of the steering wheel by equation (xxiii), then the radius of curvature of the turn becomes:

$$q = \alpha/\varphi$$

$$\omega = d(\varphi)/dt = [1/q][d(\alpha)/dt]$$

$$R = q[dt/d(\alpha)][v^2 + u^2]^{1/2}$$

$$R = [q/\tau][v^2 + u^2]^{1/2}$$

where

$\tau = d(\alpha)/dt$, the turning rate of the steering wheel.

Thus, a radius of curvature exists only when the steering wheel is being turned. When it is held at one position without moving, the vehicle simply moves in a straight line, at a diagonal to the longitudinal axis of the vehicle.

Problem 3

What is the individual loading on the left or right side wheels of the vehicle when it is executing a right curve on flat ground?

Solution

At static conditions, the following holds:

$$W = P + Q$$

where

W = total weight of vehicle,
P = load on the right side, and
Q = load on the left side.

When the vehicle turns, a centripetal force is developed:

$$F = m\, v^2/R$$

The moment equation about the left side is then:

$$P\,x - W(x/2) + Fy = 0$$

where

x = width of the vehicle, and
y = height of the center of gravity.

Solving for the load on the right side gives:

$$P\,x - \tfrac{1}{2}\,W(x) + m\,v^2/R = 0$$

$$P = \tfrac{1}{2}\,(W) - (m\,v^2)/(x\,R)$$

The load on the left side is then:

$$Q = W - P$$

$$Q = W - \tfrac{1}{2}\,W + (mv^2)/(xR)$$

$$Q = \tfrac{1}{2}\,W + (mv^2)/(xR)$$

As the equations show, as the velocity increases in the right turn, the load on the left side increases and the load on the right side decreases. This load shifting would cause the left side to "squat" down, and the right side to rise up, depending on the type and condition of the suspension system and the tires.

The above also points out that if a vehicle is turning and applies the brakes at the same time, skid marks are more likely to be made, or be heavier, on the side that is "squatting." Thus, it is not unusual for a left side skid mark to be made by a car executing a right turn, and vice versa. As noted in Chapter 2, there is also a tendency during braking to "squat" on the front wheels. Thus, when braking occurs during a turn, the following is likely to be observed:

1. If the turn is a right turn and there is braking, the left side tires will make heavier skid marks than the right side (and vice versa).
2. If the turn is a right turn and there is braking, the front left tire will leave a heavier skid mark than the rear left tire (and vice versa).
3. If there is no turn, the front tires will leave heavier skid marks than the rear tires.

The above, of course, assumes that: the vehicle has conventional front end steering; the vehicle has two axles; that the vehicle load is roughly divided in half between the right and left side tires; and that the front axle carries slightly more static load than the rear axle.

Problem 4

If a motorcyclist can lean 25 degrees to make a turn, what is the risk that the motorcycle would slide out of the turn at that angle of lean?

Solution

The general equation for the angle of lean in a turn is:

$$(m\,v^2/R) = W(\tan\theta)$$

The general equation for when slide occurs is:

$$m\,v^2/R > W\,f$$

Substituting gives the following inequality:

$$W(\tan\theta) > W\,f$$

$$\tan\theta > f$$

The above indicates the condition for when the motorcycle will slide. Slide occurs when the tangent of the lean angle exceeds the coefficient of friction between the tires and the road. If the coefficient of friction is 0.65 (dry asphalt), then the angle of lean for slide to occur is greater than 33 degrees.

Since it is assumed that this motorcyclist cannot (or will not) lean more than 25 degrees from the vertical, then he or she will not lean so far as to initiate a slide. At an angle of 25 degrees, the motorcyclist would not initiate sliding unless the coefficient of friction were less than 0.47. Since a wet roadway will often reduce the coefficient of friction by 25%, a 25-degree lean could precipitate a slide if the roadway were wet.

CHAPTER 12

Scene Examination and Solution Strategies

A. General

This chapter is a potpourri of tips and hints. The various topics discussed are too short to have their own chapter, but are too useful to be left out. It is like my kitchen drawer which has dozens of little useful whatnots all piled into one place. Like my kitchen drawer, I have made no particular effort to organize the stuff except that topics pertaining to scene examination have been listed towards the front, and topics pertaining to ways of figuring out what happened have been put in the back. (This is sort of like big stuff in the back and roll-around stuff in the front.)

In one sense, however, this is probably as good a place as any to put these topics. Some of them require that the reader be familiar with material contained in the earlier chapters. Some may not seem particularly important to the reader at first if he or she has had no first-hand experience in accident reconstruction. I can assure the reader, however, that they are all relevant, and will likely be encountered at one time or another in practice.

Lastly, I want to mildly caution the reader. Some of the topics in this chapter may seem macabre or insensitive, especially in light of the analytical approach employed in the previous chapters. They are not intended to be.

Accident reconstruction is not a "clean" discipline. In previous chapters, the discussion of scientific principles and the engineering style approach to the subject matter may have led the reader to assume that accident reconstruction is like a class in thermodynamics, or a design job. It is not. In some cases, people just like you and me have been killed or horribly maimed. It can be unpleasant to examine the interior of a car involved in such an accident. There may be civil or even criminal litigation pending. No matter what the engineer finds or concludes, it is likely that one side or the other will dispute and attempt to discredit the engineer's findings, or even the engineer personally. It is a "messy" discipline. An accident reconstruction is not simply a text book problem to solve; it is a serious matter involving people's lives, reputations, and fortunes.

B. Vehicle Contents

It is advisable that a person not reach into places in a car before visually checking them. People who use drugs will often hide their syringes under seats, under mats, in car pockets, or other places where they cannot be easily seen. It would be most unfortunate to be accidentally stuck and penetrated by a contaminated syringe.

When discovered, these items should be reported to the police agency investigating the accident. Since the police usually check a vehicle for such things after an accident, if such an item is found, it is likely that the police missed it. (If they had found it, they would have kept it for evidence.)

The finding of such items may precipitate additional testing to determine if the driver was under the influence of drugs at the time of the accident. Even if the driver was been tested for drugs after the accident, since there are so many varieties of illegal drugs, it is possible that the driver may not have been given the right test. A trace sample of the drug may tell the authorities which test to administer.

In cases where prescription drug containers are found in the vehicle, they may indicate that the driver was under medication at the time of the accident. Check the label for warnings about drowsiness, or admonitions about not driving. A count of the remaining pills compared to the issue date of the prescription may indicate whether the medication was being taken at the time. The name of the pharmacy is useful; it may be contacted by authorities to develop a medical profile of the driver prior to the accident.

Some vehicles have weapons hidden in them, especially around the driver's area. Guns, electric stun guns, crow bars, knives, chains, long metal flashlights, and baseball bats are the usual items. Martial arts items, such a numchucks, throwing stars, and darts are popular with people who belong to karate clubs. Mace, pepper gas, stun guns, and amazingly loud personal protection alarms are also common.

In rural areas where pickup trucks are popular, weapons such as shotguns, rifles, cross bows, and compound bows are often mounted in plain sight over the rear window, or stashed behind between the seat and the gas tank for easy access. Occasionally, a cattle prod may be found under or behind the seat.

The above weapons are mentioned because some of them may cause injury to the person who accidentally finds them. Pity the person who inadvertently trips a personal protection alarm.

The presence of spilled food items in the car, especially around the driver's area, might indicate that at the time of the accident the driver was distracted by eating or drinking. Maps laid out on the front passenger seat may also indicate that the driver was distracted. There have been reported instances of portable television sets being found on the dashboard with the switch in the "on" position. Some regular expressway drivers have even been known to read books or business reports while driving.

The side car pockets or the glove compartment will often contain a copy of the car owner's manual. This is handy to locate fuses, or determine where certain features of the vehicle are and how they work. It will also contain performance data, maintenance recommendations, warnings, precautionary notices, and operational instructions.

The driver's door often will have oil change stickers on the inside. Sometimes the dates on these stickers are useful in tracking down maintenance information. Often, the sticker will have a date, and the mileage on that date. This can be useful in making estimates of car usage over various periods of time.

Any items removed from a vehicle must be documented. When criminal proceedings are pending, the appropriate law enforcement authorities must be notified if any items are removed, and what will be done with them. Usually, evidential chain of custody type documentation will be required. It goes without saying that any items that have significant worth, like a diamond earring, must be reported and returned to the proper owner. Photograph all items of importance that are found. Despite everyone's best efforts, important evidence does get lost when cases sometimes take several years to reach trial.

C. VIN

The VIN (Vehicle Identification Number) is usually located at several points in a car, most notably at the base of the windshield, left hand corner of the driver's side dashboard, and also on the inside of the driver's side door. The VIN number can be used to track the maintenance records of the vehicle through any convenient dealership, since most of the major car companies maintain centralized computer data banks into which the dealerships deposit maintenance records, etc. Recall notices specific for that make and model are also often coded by VIN number.

With a vehicle's VIN it is possible to:

- determine whether it has been brought back to the dealership to have a recall item fixed,
- determine the maintenance records of the vehicle through authorized dealerships,
- determine at which dealerships the maintenance was done.

Coded into the VIN number is the vehicle type, the factory where it was made, and the year and week it was built. Most car companies currently encode the VIN in a bar code so that it can be read by a scanner from the outside of the windshield.

Figure 1. Spider cracking of windshield due to impact by driver's head. Hair strands were found in crack.

D. Who Was Driving?

An interesting question that comes up now and then is who was driving at the time of the accident. This question can be particularly important when there is only one survivor. It may be that the estates and families of the deceased occupants may wish to sue the driver for wrongful death, negligence, or similar. It may be that the survivor was found to be very drunk after the accident, and is subject to criminal prosecution if he was the driver. Thus, there can be a lot of reasons why a survivor may not wish to be identified as the driver.

In some instances, there may be a person who claims to have been the driver, but was not. Perhaps the person who actually was driving has some legal reason why he should not have been operating a vehicle, and a loyal buddy or family member is attempting to "cover" for him.

In a bizzare case in which I was involved, there was a one-car accident at night where the driver was drunk, and his car went off the road and crashed through the garage door of a car repair shop. The driver, who had already been convicted of several DUI's and had lost his license, jumped out of the car and ran home, which was just a couple of blocks away. The driver then called the police and reported that his car had been stolen while he had been asleep.

Meanwhile, the sound of the crash prompted several neighbors to call the police. When the police arrived at the garage, it took them a while to sort out what had occurred. While the garage door had obviously been smashed in, the smashed car had come to rest in a repair bay with several other similarly smashed cars. Even the garage owner was not sure at first which was the rogue wreck, since newly wrecked cars were occasionally hauled into the repair bay after closing hours by a wrecker service. And, of course, there was no driver around to question.

While the wrecked cars were being sorted out at the garage, two other officers were interviewing the owner of the "stolen" car. It was apparent to the officers that the owner was drunk as he talked during the interview. They also noticed that he had a broken arm and was bleeding across his face. Since this same person had reported that he had been asleep all night, it was difficult to convincingly explain how he had gotten drunk, seriously cut his head, and broken his arm while asleep in bed. The position of the cut on his head matched the location of a "spider" type impact crack on the windshield, and the location of the break in his arm matched the location of an obvious bend in the steering wheel. The car owner was eventually convicted of another DUI and was required to serve time.

When the occupants of a car do not wear seat belts, the identification of the driver is generally easy. Often at impact, the driver will be hurled against or through the windshield in the direction of impact. If the driver was hurled against the windshield but the windshield did not break out, it is likely that hair and tissue of the driver will have been left in the "spider" crack pattern in the windshield. If the driver was the only one in the car with blond hair, long hair, or some other unique physical quality, it can be as simple as examining the hair found in the windshield cracks under low magnification and comparing color, length, or diameter to the hair of the other occupants.

If the driver is tall with respect to the windshield, it might be that he will abrade his head along the top of the ceiling, or impact it on the sun visor. Embedded hairs can often be found there also.

Because the driver has the steering wheel in front of him, if he is not seat belted, the steering wheel will often cause unique injuries to the driver. Bruises in the shape of the steering wheel are an obvious giveaway. Often, where the steering wheel has been bent up or damaged, the driver will have corresponding injuries on his body corresponding to the same position.

If a driver is injured and bleeds in the car, the blood type of the stains can determined. This is especially useful if the driver's blood type differs from the rest of the occupants' blood types. This type of chemical analysis, however, requires the services of a specialized laboratory, which may not be readily available in some areas. Some searching may be required to find a laboratory that does this type of analysis.

The position of the other occupants of the vehicle at the time of the accident may also be determined by reviewing each person's injuries with respect to his or her trajectory through the vehicle after impact. For example, a person in the back seat may have been thrown forward between the two front seats head first into the dashboard. Thus, that person would have head and shoulder injuries consistent with that trajectory.

E. Metal Fracture

Often, the cause of a crash or accident is blamed on the failure of an important metal part. It may be claimed that the part failed first, which then precipitated the accident.

In such cases, it is important to examine the fracture and determine its metallurgical characteristics. A hardness test will determine its approximate strength and provide information as to whether or not it has been heat treated. The type of fracture is also important. A fatigue fracture indicates that the part has been partially cracked for some time prior to the accident. Thus, it can be argued that the part was weak at the time of the accident and susceptible to failure.

A shear fracture, tensile fracture, or bending fracture with no indication of fatigue usually indicates that the part failed at the time of the accident, perhaps being damaged as part of the accident process. This is especially true when the fracture surface is new, and no evidence is observed that the fractures existed prior to the accident. Evidence that the fracture or fractures may have been present prior to the accident include discoloration of a portion of the fracture from weathering, dirt, corrosion, or exposure to grease and oil. Such fracture may have been caused by previous accidents or abuse.

When a part fails due to loss of material from corrosion, usually the effects of corrosion are plain. A common type of corrosion is caused by salt from either road salt in cold climates, or from the sea in coastal climates. Less common types of corrosion can occur if the vehicle is used at a factory or place of work where it is driven into an area where it can come into contact with chemicals.

Sometimes when a car front end or whatever is reassembled after major repair work, the mechanic may have lost an original nut or washer. The nuts and bolts in a vehicle are usually carefully specified at the factory for hardness, strength, and alloy content. If the nut is simply replaced with one that "fits" instead of the factory specified replacement, it is possible that the replacement nut will not perform properly. It may back off, rust out, or simply fail due to weakness.

When a mechanical failure has occurred in a vehicle because of a fastener, it is best to check that the nut or bolt is the type called for by factory specifications. This will often require a hardness check, and perhaps even a chemical analysis of the alloy. A mass spectrometer may be used to check the elemental components of the alloy. Mass produced bolts and nuts that are graded to accepted standards have "tick" marks on the cap that indicate the strength of the alloy used in the fastener. Bootleg fasteners, however, have been known to have the same type of markings on their caps to fool the buyer. Bootleg or substandard part replacements can precipitate unexpected mechanical failures in a vehicle.

F. Improper Welding Repairs

In a similar vein, improper welding repairs may sometimes be done that eventually cause a mechanical failure and precipitate an accident. Many vehicles and trailers have frames and structural elements made of high-strength alloy steel. Such steels are often heat treated to obtain the high strengths required. If a body shop attempts to weld these high-strength alloy parts with common welding rod or wire rather than replace them, the weldment will not have the same qualities and strength as the original base metal. Further, metallurgical changes in the base metal in the heat affected zone immediately next to the weld will weaken the frame or structure. Usually, the metallurgical damage consists of "softening" of the metal. The material in the heat affected zone (HAZ) is weaker and more susceptible to fatigue failure than before.

While welding on these types of materials is, of course, possible, most car repair shops and dealerships do not have the materials nor are their welders familiar with the pre- and post-weld heat treatment procedures necessary to successfully accomplish it. Some shop welders mistakenly believe that if a person cannot visually distinguish where the weldment has been made, it is a good weldment.

The presence of a weldment on heat-treated, high-strength alloy steel, however, is easily detected with a field hardness tester. Some hardness measurements taken from the base metal are compared to those taken across the area of the failure will reveal if the base metal has been welded, and the extent of the heat affected zone next to the weld. While a welder can grind and contour a weld to the point where it may be visually indistinguishable from the rest of the material, he cannot hide the hardness of the material. If needed, small pieces of base metal and material from the weldment can be prepared for either mass spectroscopic analysis, or fractograph presentation to show the difference in alloy constituents and grain structure.

In one case, a low boy trailer had been hauling a bulldozer when the low boy suddenly dropped down on the right side and rolled over. This occurred when the tractor-trailer was at the bottom of a hill after descending a steep grade, and while also making a left curve. The action of the low boy caused the tractor to also lose control, and the bulldozer was lost and went caroming of the roadway.

Inspection of the low boy found a weldment in the frame where a broken brace had been repaired. The low boy frame was made of high strength steel H-beams, and the weldment was made of common E6010 welding rod, which is mild steel. Hardness measurements in the base metal were about 340 BHN, while those in the heat affected zone and weldment ranged from 90 to 110 BHN.

A check of the weight of the bulldozer as compared to the load rating of the low boy, as stated on the manufacturer's spec plate, found that the trailer was not overloaded. However, an analysis of the stresses where the weldment had been made found that the new weldment could not carry the combined load of its own weight, the weight of the bulldozer, and the additional load caused by two kinds of "mv^2/R" forces: one caused by "pulling up" at the bottom of the hill which adds to the downward forces all around, and one caused by the left turn curve which caused load shift to the right side of the trailer.

G. Brake Fluid Compatibility

In some older foreign cars, using the wrong kind of brake fluid could precipitate an accident. The older European cars often used natural rubber for seals and o-rings in brake systems. The U.S. type cars used neoprene type synthetic rubber for the same parts. The brake fluid used by U.S. cars would actually dissolve the natural rubber-type parts, causing them to leak and fail. Thus, if a person made a mistake and did not put in the appropriately compatible brake fluid, a brake failure could occur and possibly precipitate an accident.

H. Tires

Tires often are blamed for accidents. The most common type of tire failure occurs when a tire is run underinflated. When this happens, the bending stresses in the tire wall are greatly increased, and the tire will heat up. Since the strength of the rubber and carcass of a tire is sensitive to temperature, the tire wall will have less mechanical strength at the elevated temperature. The combination of high stresses and heat build up causes the materials in the tire to fatigue. Eventually the tire will blow out.

A tire that blows out due to underinflation will usually have a blow out in the tire wall. It often will be located at the point of maximum bending in the tire wall. If a tire has been run underinflated for a time, the point of maximum bending can often be clearly observed by discoloration in the tire wall. There will be a boundary line that runs around the tire at a constant radius from the center where there is darkening on one side of the line, and normal tire color on the other.

The "hardness" of a tire or other rubber product is measured with a durometer. If a tire has been run underinflated, the durometer number of the heat affected tire wall will be measurably lower than that in areas that were not subject to the excessive bending stress and heat build up. Thus, a "hardness" profile of the tire can be taken to map out the extent of softening that had occurred prior to the blow out.

Also, when the tire blows out due to underinflation, the point where the blow out occurs will exhibit material fatigue. The cord fibers which first failed will be frayed unevenly and will have a disheveled, even "dirty" appearance. This will be in contrast to fibers which simply broke apart in tension later on when the tire failed generally. Also, where the bending stresses were highest in the tire wall, usually at a point closest to the exterior surface, the fibers will have come loose from the rubber matrix and may have even "wallowed" around in the matrix. This is in comparison to the fibers in the rest of the tire which may still adhere fast to the matrix.

One particular type of tire that seems to be run underinflated often is the racing type, low-aspect tire. This is usually a radial type

Figure 2. Tire failure due to sidewall impact with a sharp object.

tire that is wide but is not very thick in the sense that the wheel rim sits very close to the ground even when the tire is properly inflated. These tires appear nearly the same when properly inflated as they do when underinflated. Thus, a simple visual inspection is not a good way of determining if the air pressure in them is adequate. Also, they usually require about 30–50% more air pressure than typical car tires. A person used to normal car tire pressures of 26–32 psi may not know that these tires require 38–50 psi inflation pressure.

When a tire fails a considerable time before the accident, the tire will often be damaged by having been run flat. There may be cuts in the tire from the rim, and even holes where the pavement abraded away material if the brakes were locked. If the wheel continued to roll some time after the tire lost air, the rims may be damaged, and rim marks may be visible in the pavement. The presence of rim marks in the accident trajectory of the vehicle can be used to determine that the tire was flat.

Impacts to the tires by curbs, pot holes, and the like can be severe enough to cause a tire to lose air and deflate. Often, the rim will be bent where the impact occurred, and this will also correspond to tire damage at the bead. Since most late-model cars and trucks use tubeless tires, damage to the tire bead by impact will usually cause the tire lose seal and deflate.

Severe turns can sometimes cause a tire to slip off the bead seat of the wheel and go flat. When this occurs during a curve or turn, the tire edge or scuff marks will become darkest just before the tire fails. Then rim marks may be observed in the pavement.

Examination of the tire tread can also provide clues as to whether or not the tire was consistently run with low or with high pressure. A consistently underinflated tire will have greater tread depth in the middle than at the edges. Conversely, a consistently overinflated tire will have less tread depth in the middle than at the edges.

Figure 3. Note tire rim damage. Tire ran flat before impacting the curb.

In older type "recap" tires, air would sometimes escape from the older portion of the tire and diffuse into the layer between the old tire and the newer recapped tread. This would cause a lump or carbuncle to form on the tire which would eventually burst. Such tires would usually run rough just before they failed. Drivers would describe it as feeling like driving with a brick strapped to the tire.

If a tubeless tire is not mounted properly, it is possible for the tire to have a "hung bead." This is where the bead of the tire is not properly seated on the bead seat of the wheel. However, it may be sufficiently seated to hold air during initial mounting and low-speed driving. The tire may suddenly fail during cornering, making a curve at higher road speeds, or simply encountering typical bumps in the roadway. An excellent study of this mode of failure is contained in the publication, "Research Dynamics of Vehicle Tires, Volume 4" by Andrew White. While I understand this document is out of print, it is still available to read through library loans. (See the "References" chapter at the end of this book.)

Naturally enough, when a tire fails due to punctures from something on the road like a nail or sharp object, the puncture will occur in the tread portion of the tire. Dismounting of the tire and inspecting the tire interior may occasionally find the item responsible for the puncture embedded within the tire itself. Damage from a sliding contact with a curb or curb offset will usually cause damage to the outer side wall of the tire. Tire failure due to physical damage by impact or contact will usually be apparently from physical examination of the tire carcass. The rubber in the vicinity of the damage will often be gouged inward or similar.

One tire-related type of accident involves pavement drop offs. Some road shoulders may be lower than the roadway. When the difference in elevation is more than 2–3 inches, and a vehicle goes off the roadway, getting back on may be a problem. The tire may not be able to climb back onto the roadway due to the height difference. This is especially true if the drop off is straight rather than tapered. The tire may scuff along the edge of the roadway, abrading the inside of the tire. This can cause loss of control of the car.

The proper way to get back on the roadway in this case is to slow down, and then turn sharply back onto the roadway. It is necessary to slow down because getting back on the roadway at moderate or high speeds may cause the car to overshoot into the other lane. It is necessary to turn sharply because a small angle may not be able to "catch" the edge properly and provide sufficient traction to lift the front end back onto the roadway. If the angle is too small, the tire will momentarily rise up and fall down the edge of the roadway, producing erratic bumping and tire forces.

Figure 4. Low-aspect tire which failed from operation at low pressure. Note discoloration due to overheating.

I. Point of Impact

This topic is probably one of the most contested items associated with scene examination. In fact, in some states it is not permitted to discuss in court the point of impact or even use the term due to questionable past practices in accident reconstruction employing what was called the debris method.

The debris method is simply an examination for debris on the ground at the accident scene like headlight glass, turn signal lens plastic pieces, and similar objects which were damaged by impact between the two vehicles. The general area where these items were found would be mapped out, and the visually determined centroid of that area was considered to be the point of impact of the collision.

This is a very poor method because it does not take into account the difference in speeds between an impacting vehicle and the items separated from the vehicle by impact. The debris items may actually travel and come to rest several yards away from where impact occurs.

Sometimes, when a collision has damaged the radiator or engine, the position of spilled radiator fluid or motor oil would be called the point of impact. This is also a poor method, because it may take some time for either fluid to actually drip down onto the pavement. During that time, the vehicles could have moved considerably away from the actual point of impact. While the positions of the spilled fluids do mark the location of the vehicles after they came to rest, they do not generally mark the point of impact.

A more accurate method is to examine the scene for tell-tale tire marks or gouges. A skid, rim or other tire mark that suddenly deviates off in a new direction marks a point of impact. As noted in the Example Problem 3 in Chapter 10, a severe head-on collision will sometimes cause the front tires to become flat due to downward forces. Thus, a skid mark that terminates at a short rim mark or rim gouge will denote the point of impact.

Similarly, when impact occurs, other parts of the car may be damaged and be pushed down to the pavement making gouges at or leading away from the point of impact. Tie rods and bumpers are particularly prone to doing this.

J. Weigh Stations

When analyzing accidents involving trucks, weigh stations can sometimes provide very useful data. When a commercial truck is checked at a weigh station not only is its weight checked, but the time it was checked is often stamped on the weigh ticket. This places the truck at a known location at a known time.

This is useful if the truck is then involved in an accident after leaving the weigh station and it is suspected the truck driver was speeding. When an accident is reported, the time is recorded by the dispatcher. Further, when the officer arrives at the scene, he usually checks and determines how long before the report of the accident was made that it occurred.

Thus, combining the time and location from the accident report and a weigh station ticket provides an accurate base for determining the average rate of speed of the truck between the two points. Obviously, if the average speed is above the speed limit, there must have been more distance driven above the speed limit than below it. This is especially useful if the driver of the truck has claimed during the accident scene interview that he drove the whole distance from the weigh station to the accident scene at or under the speed limit.

K. Alcohol Content of Blood

In some states the amount of alcohol in a person's blood may not exceed 0.10% or that person is considered legally drunk. Recently, some states have adopted a standard of 0.08%. In a few states, the legal limit may be as low as 0.06%. In any case, the use of alcohol is known to reduce the reaction time, awareness, and concentration of a driver. In most states the refusal to take an alcohol blood or breath test is considered an automatic admission of drunkenness, although there have been some significant court challenges on the basis of the Fifth Amendment.

Sometimes a driver cannot be tested for alcohol right away. Due to various delays, the test may be administered several hours after the accident. If the driver of the car involved in the accident is somewhat savvy in these matters, the delays may be aided by his own actions, knowing that as more time passes, he will have less alcohol in his blood as his body metabolizes it.

In one case, the driver of a car involved in an accident, who was also suspected of being under the influence, walked over to a nearby tavern to ostensibly call in the accident. While at the tavern, and while waiting for the police to show up, he ordered several shots of hard liquor from the bartender and in front of other witnesses. When the police showed up and breath tested him, he did not pass. However, he was successfully able to claim that the alcohol was from the drinks he took in the tavern, in front of witnesses and clearly after the accident, rather than from drinks consumed prior to the accident. The reason that it was suspected he had been drunk at the time of the accident was that his driving was erratic, and he had bypassed several other telephones along the way to get to the one at the tavern.

The amount of alcohol in a person's blood decreases more or less linearly over time as the alcohol is metabolized. The following equation can be used to estimate a person's blood alcohol level at a time previous to the testing:

$$X_1 - m\,t = X_2 \tag{i}$$

where

X_1 = % alcohol in blood at previous time,
X_2 = % alcohol in blood at time tested,
t = time in hours, and
m = hourly rate of metabolic consumption of alcohol from the blood.

In general a rate of 0.017%/hr is used for men, and a rate of 0.015%/hr is used for women. Thus, a delay of one hour in having a breath test done can reduce a man's blood alcohol level from say 0.105% to 0.088%, the latter being below the legal limit in many states.

If the number of drinks a person consumed before the accident can be determined, the blood alcohol level just after consuming those drinks can be estimated from body weight versus number of drink type tables. If a breath test or blood alcohol test can be done at a specific time after the accident, a graph of the blood alcohol percentage versus time can be constructed by connecting the two known points by a line. With this graph, an estimate of the blood alcohol at the time of the accident can be made rather accurately, which even accounts for the individual's variations of metabolism from the average.

L. Presumption of Sobriety

When a fatal accident involves only one car and driver, some investigating agencies will often not bother to have the deceased checked for either alcohol or drugs. It is reasoned that there is no one alive to prosecute, and such tests incur costs to law enforcement agencies which are already tightly budgeted. However, this apparently reasonable cost saving omission can cause difficult problems in the civil law arena.

Often, in making a reconstruction, the engineer may suspect that a driver was drunk, drugged, or sleepy by the lack of appropriate action during the accident trajectory. For example, suppose that there is a slight curve in a roadway, and that the vehicle simply went straight instead of turning with the curve. If the car had no problems, was not speeding, and there is no evidence that either braking or evasive turning was done by the driver, it is reasonable to suspect that the driver was "out to lunch" while things were occurring. This is especially true if the driver had plenty of time and space in which to act.

However, if no drug or alcohol test was done on the deceased, there is a presumption of sobriety. Without the appropriate tests, the matter cannot be firmly established. It might be argued by the lawyer for the deceased that the driver's inaction was due to confusion or fear rather than drugs or alcohol.

Thus, if the estate of the deceased sues the city or state for a dangerous roadway condition such as insufficient signage, or sues the manufacturer of the car for some type of latent defect which caused the accident, the defense has no specific evidence to establish that the driver was simply not functioning at the time of the accident. It can only be inferred from the omission of reasonable actions which would normally have been taken by an alert driver. Thus, the defense is put through the time and expense of proving that everything else was all right. This, of course, means that it is worth something to the defense to settle the matter out of court rather than to assume all the legal expenses, engineers, etc., and then risk a jury trial.

M. Setting Limits

If the reader can remember back to the Chapter 1, the concept of setting limits was introduced in Section F. While an exact solution to an accident problem may not be possible, it may be possible to partially solve the problem as an inequality.

For example, suppose that a passenger sedan travels along a thoroughfare marked for 40 mph. At some point, a pickup truck pulls out from a side road in front of the sedan, and the two vehicles collide. Prior to contact, the sedan skidded 40 feet. After contact, both vehicles skidded 15 feet more in the general direction of travel of the sedan. Because of the accident occurred some time ago, the vehicles are no longer available for examination, and an examination of the accident scene has long since lost any value. Thus, an estimate of energy dissipated by crush can not be determined.

However, despite these problems, an inequality can be set up. Assuming a COF of 0.6, a lower limit speed for the sedan can be calculated:

$$v = [2(32.17\ \text{ft/sec-sec})(0.6)(40\ \text{ft} + 15\ \text{ft})]^{1/2} = 46\ \text{ft/sec or } 31\ \text{mph}$$

From this, it is known that the sedan must have been travelling at or more than 31 mph when skidding was initiated. Since the speed of both vehicles just after impact was 26 ft/sec, then, an estimate of the speed of the can be made by making a momentum balance:

$$m_{sedan}v_1 + m_{truck}u_1 = (m_{sedan} + m_{truck})v_2$$

$$v_1 = [(m_{sedan} + m_{truck})v_2 - m_{truck}u_1] / m_{sedan}$$

$$v_1 = [(1 + mr)v_2 - (mr)u_1]$$

where

$$mr = m_{truck} / m_{sedan}$$

However, the forward speed of the pickup, "u_1," at the time of impact is not known.

Despite the fact that the forward speed of the pickup is not known, it is known that the speed would have been relatively low. This is because the pickup was just pulling out from a stop and was making a left turn while carrying a load. This means that in the above equation, the term "$(1 + mr)v_2$" sets a lower limit to the speed of the sedan because the "$(mr)u_1$" term can only increase the answer for "v_1" (the sign of "u_1" is opposite of that of "v_1").

Thus, if the light-weight pickup weighed 25% as much as the luxury sedan, then the lower limit to the speed of the sedan at impact is:

$$v_1 > [(1 + 0.25)(26 \text{ ft/sec})]$$

$$v_1 > 32.5 \text{ ft/sec or } 22 \text{ mph.}$$

If it is assumed that the pickup truck accelerated in a normal fashion from the point it was stopped to the point of impact, it can be reasonably assumed that the truck would have had no more than a 30 ft/sec and no less than a 20 ft/sec forward speed. Thus, the "$(mr)u_1$" term can only add between 5 and 7.5 ft/sec to the lower limit already figured. In one sense, the term "$(mr)u_1$" is relatively insensitive to the forward speed of the pickup. A 10 ft/sec difference in pickup truck speed only makes a 2.5 ft/sec difference in the calculation of the speed of the sedan.

Thus, the momentum equation indicates that the sedan was traveling between 37.5 and 40 ft/sec at impact. If we then add to that the speed indicated by the 40 feet of skid prior to impact, then the speed of the sedan just prior to skidding is bracketed as follows:

$$(1544.16 + 1406.25) < v_1^2 < (1544.16 + 1600)$$

$54.3 \text{ ft/sec} < v_1 < 56.1 \text{ ft/sec}.$

Thus, despite the fact that we did not have enough information to arrive at an exact answer, the bracketing is tight enough that it is very close to an exact answer. In fact, the bracketing is small enough where it can be concluded that the sedan was not exceeding the speed limit of 40 mph when skidding initiated.

N. Accident Avoidance Strategy

In this method, it is assumed that certain parties to the accident travel at the legal speed, and perform reasonably in all their actions, including braking, turning, and accelerating. The result is then compared to what actually occurred. If there is a difference, then this difference lends itself to an explanation of what occurred.

For example, suppose that a motorcyclist was travelling down the road when he saw a large truck coming in his direction make a left turn in front of him. The motorcyclist skids 48 feet and then lays the bike down. While he never contacts the truck, the rider claims that the truck turned in front of him causing him to have the accident.

It is known where the bike came to rest after being laid down, and it is known where the truck was located while making the left turn. In this case, the bike came to rest 15 feet from the path of the truck, and the end of the skid mark was located 80 feet from the path of the truck. The posted speed limit was 35 mph.

If the motorcycle had been travelling at 35 mph, and had begun braking where the skidmark started, but braked in a normal fashion using a deceleration of 16 ft/sec^2, then the distance to come to a complete stop from 35 mph would be:

$$d = [(51.33 \text{ ft/sec})^2] / [2(16 \text{ ft/sec}^2)] = 82.3 \text{ ft}$$

It is known that the skid mark started 128 feet from the path of the truck, and it is known that a normal stop from the speed limit would only take 82.3 ft. Because reaction time occurred before the initiation of the skid mark, it is not necessary to consider it in the above. The above information then indicates either:

- the motorcycle was excessively speeding, or
- the motorcyclist lost control of the bike due to lack of skill, panic or other reason.

Figure 5. A "triple." It is alleged that a quick set of left-right-left turns can cause a "crack the whip" effect in these vehicles.

A lower-bound estimate of the speed of the motorcycle from just the 48 feet of skid marks using a COF of 0.7 finds that it was going at least 46 ft/sec or 32 mph when skidding began. Since the motorcycle then slid another 65 feet on its side, an estimate of the energy consumed by side sliding can be made assuming a COF of 0.25. This then gives the speed of the motorcycle as 57 ft/sec or 39 mph, which is slightly over the speed limit.

But even at 39 mph, the motorcycle could have easily come to a complete stop with typical braking within 128 feet as the following shows:

$$d = [56.6 \text{ ft/sec}]^2 \;/\; 2[16 \text{ ft/sec}^2] = 100 \text{ ft}.$$

Thus, it is concluded that the motorcyclist lost control of his vehicle since he had plenty of time and space to avoid colliding with the truck, even though it may have turned in front of him. In fact, the motorcycle did not even have to come to a complete stop to avoid the contact with the truck; it only needed to slow down enough to give the truck time to complete its left turn. Thus, the question now becomes, did the truck's actions cause the motorcycle accident, or was it the lack of skill of the rider?

Most states require that motorists exercise defensive driving and accident avoidance. Thus, if car A were to pull out in front of car B, the person in car B does not have the "right" to clobber car A even if the driver of car A has committed a mistake. A person has an obligation to avoid accidents by reasonable and prudent defensive actions.

O. Air Bags

A useful tidbit to know about air bags is that they are set to operate when the deceleration forces on a vehicle are equivalent to that of a 12-mph front impact with a fixed barrier. This is particularly handy, since an air bag is actuated by a calibrated accelerometer. When assessing low to moderate speed impacts, the fact that an air bag operated may provide a firm lower limit for deceleration.

CHAPTER 13

References

A. General

The reconstruction of vehicular accidents may be considered by some to be a somewhat arcane science. Unless a person knows where to begin looking for informational resources, most libraries will appear barren of related materials.

The following bibliography lists some reference materials that are useful in the analysis of vehicular accidents. Where appropriate, I have added comments to let the reader know what the item contains. There is no particular significance to the order of items listed, and the list is certainly not to be considered exhaustive. I have emphasized books and bound collections of papers more than single papers, because the former are more easily obtained by inter-library loan, and provide more "stuff" for less effort.

B. Books

Motor Vehicle Accident Reconstruction and Cause Analysis, by Rudolf Limpert, Second Edition, 1984, The Michie Company, Charlottesville, Virginia. LCCC No. 78-62023, ISBN 0-87215-646-X.

Despite the fact that this text is published as one of a series of litigation-related texts for lawyers, it is a first-class textbook for accident reconstruction. The approach is technically sound and is mathematically relatively sophisticated when appropriate.

The Traffic Accident Investigation Manual, by Baker and Fricke, Ninth Edition, 1986, Northwestern University Traffic Institute, LCCC No. 86-606-16, ISBN 0-912642-06-8.

This is an often referenced text used to train police officers. The emphasis is on collecting and preserving accident scene information. The accident analysis sections are rudimentary but sound. The qualitative information contained in the text is very good. It is a good idea to be acquainted with this text since it is so commonly referred to in court proceedings, police reports, and so on.

The Way Things Work: An Illustrated Encyclopedia of Technology, 1967, Simon and Schuster, New York, LCCC No. 67-27972.

This is a handy compendium of short articles and explanatory diagrams of many appliances and machines. With respect to accident reconstruction, there are many fine articles and illustrations in the several volumes that deal with automotive brakes, steering systems, transmissions, and so on.

The Way Things Work, by David Macaulay, 1988, Houghton Mifflin, ISBN 0-395-42857-2.

The sections about friction, clutches, brakes, and synchromesh transmissions have very instructive illustrations.

Field Accidents: Data Collection, Analysis, Methodologies, and Crash Injury Reconstructions, Society of Automotive Engineers, February 1985, ISBN 0-89883-720-0.

One of several in a series published by SAE. This is an excellent collection of papers dealing with accident reconstruction. Article 850437 is especially useful with respect to crush stiffness coefficients.

Accident Reconstruction: Automobiles, Tractor-Semitrailers, Motorcycles and Pedestrians, Society of Automotive Engineers, February 1987, ISBN 0-89883-454-6.

This is a collection of papers representative of the direction of research in this field in 1987. I recommend two papers on braking, 870501 and 870502.

Work Zone Traffic Control, Standards and Guildlines, U.S.D.O.T., Federal Highway Administration, 1985, also A.N.S.I. D6.1-1978.

This booklet describes the standards for traffic control in roadway construction areas. Such information is a must in assessing vehicular accidents occuring in construction areas.

An Examination of Sudden Acceleration, U.S.D.O.T., National Highway Traffic Safety Administration, Report Number DOT-HS-807-367, January 1989.

This study exhaustively examines sudden acceleration incidents (SAI's) which include stuck accelerator cases, runaway car engines, and so on. The report contains information relating to the famous Audi accelerator case and other useful test data.

1985 SAE Handbook, Volume 4: On-Highway Vehicles and Off-Highway Machinery, Society of Automotive Engineers, ISBN 0-89883-867-3, LCCC No. 25-16527.

This handbook is regularly updated from year to year and contains a wealth of information with respect to accident reconstruction. Standard test procedures, performance standards, and design specifications relating to vehicles and equipment are covered. It is fundamentally important to be familiar with this document.

Statistical Abstract of the United States, U.S. Printing Office.

This collection of facts and figures is published annually by the U.S. Census Bureau. Within it are hoards of information relating to traffic accidents, registered vehicles, road miles driven, and so on.

Atlas of Stress Strain Curves, edited by Howard Boxer, ASM International, Metals Park, Ohio, 1987.

This is a useful source of information about metals when a person is calculating expended crush energy. Stress-strain pull test charts for most of the alloys used in automobiles and vehicles are listed.

Mathcad, Version 2.5, Addison-Wesley, 1990.

While not strictly a "book", this software is extremely useful in working out the equations concerning accident reconstruction. An interesting sample problem concerning braking is included in some sample files under BRAKE.MCD. An iterative energy method is employed to obtain a plot of remaining kinetic energy versus time during stopping. The sample problem is well worth noting. Currently, there is also a new "Windows" type version of Mathcad available. It comes with several handy built-in engineering handbooks.

Introduction to Forensic Engineering, by Randall Noon, P.E., CRC Press, ISBN 0-8493-8102-9, 1992.

This book broadly discusses the investigation of a number of accident events including car accidents on an introductory level. Other chapters in the book, especially courtroom procedures and the application of the scientific method to accident reconstruction are well worth perusing, if I do say so myself.

Manual on Uniform Traffic Control Devices for Streets and Highways, National Joint Committee on Uniform Traffic Control Devices, U.S. Department of Commerce, Bureau of Public Roads, June 1961.

This is the "Bible" of traffic signs, traffic lights, sign and light placement, road markings, and general signage requirements.

Symposium on Vehicle Crashworthiness Including Impact Biomechanics, edited by Tong, Ni, and Lantz, American Society of Mechanical Engineers, AMD-Vol. 79, BED-Vol. 1, 1986.

This is a collection of papers presented at the ASME Winter Annual Meeting in 1986. With respect to automotive crashes, the paper which begins on page 91 is particularly useful in its discussion of barrier collisions.

Galileo's Revenge - Junk Science in the Courtroom, by Peter W. Huber, Basic Books, 1991.

While not a technical book, this is very interesting reading from the standpoint of liability law. It discusses the apparent proliferation of "junk science" testimony in the court room. Chapter 4 is especially interesting as it deals with sudden acceleration incident lawsuits.

I recommend reading chapter 4 of Mr. Huber's book after reading chapter 8 of this text to compare the legal maneuvering with the technical underpinnings of the subject.

Bicycle Accident Reconstruction: A Guide for the Attorney and Forensic Engineer, by James Green, P.E., Second Edition, Lawyers and Judges Publishing Company, 1992.

This is one of just a handful of publications that deal with vehicular/bicycle accidents. The book seems to be directed more to lawyers than engineers. It also repeatedly extols the virtues of qualified bicycle mechanics to the point where the reader might come to believe that no one but a bicycle mechanic "really" understands how a bicycle works. Despite this bias, however, the book has useful information about the principle of conspicuity as applied to bicyclists. Some of this information is applicable to motorcyclists.

Basic Principles and Laws of Mechanics, by Alfred Zajac, D.C. Heath and Company, 1966.

This is volume I of a two-volume textbook set on physics for the first year undergraduate or advanced undergraduate. Chapters 8, 9, and 10 provide an excellent introduction into the use of Lagrangian equations with generalized coordinates. While there are many texts that ably discuss Lagrangian equations with generalized coordinate systems, I personally like this one because the notation is easy to follow.

Energy Methods in Applied Mechanics, by Henry Langhaar, John Wiley and Sons, Inc., 1962.

Langhaar's text has been a mainstay in mechanicals for some time. Section 6.2 gives a good derivation of Euler's columnar buckling formula, and further discusses post buckling behavior.

Formulas for Stress and Strain, by Raymond Roarck, 1943, McGraw-Hill.

While I realize that much newer editions are available, I personally prefer the layout of the older versions of this venerable reference. In fact, a software version is currently available to go with the *T-K Solver* program.

The text has excellent sections on columnar buckling, and plates and shells buckling. Table XVI, item M shows a thin-walled cylinder under uniform longitudinal compression, and the shell shown in item K in the same table is very similar to a car door panel under undergoing compression from a frontal impact.

Research Dynamics of Vehicle Tires, Volume 4, by Andrew White, Research Center of Motor Vehicle Research of New Hampshire, Lee, New Hampshire, 1965.

While this book is out of print, copies are still available through libraries. This particular volume of the series is about improper mounting of tires, and the hazards associated with it. It may be the best all-around discussion of this subject in the literature. I was especially impressed with the photographs showing what a tire can do when it explodes during mounting on a wheel rim.

C. Papers

Detroit News, *Five part series of articles on SAI's*, December 13–17, 1987.

Unintended Acceleration by Tom Lankard, Autoweek, January 19, 1987.

The Occurence of Accelerator and Brake Pedal Actuation Errors During Simulated Driving, by S. B. Rogers and W. Wierwille, Human Factors, 1988, 30(1), pages 71–81.

The Amateur Scientist by Jearl Walker, Scientific American, February 1989, pages 104–107.

Excellent article which discusses modulated braking and braking strategies to minimize braking distance.

The Amateur Scientist by Jearl Walker, Scientific American, August 1989, pages 98–101.

Interesting article which discusses the wave motion of highway traffic flow.

The Physics of Traffic Accidents by Peter Knight, Physics Education, Volume 10, No. 1, January 1975, pages 30–35.

An Overview of the Way EDSMAC Computes Delta-V by Day and Hargens; Engineering Dynamics Corp., SAE 880069.

Differences Between EDCRASH and CRASH3 by Day and Hargens, Engineering Dynamics Corp., SAE 850253.

Energy Basis for Collision Severity by K. L. Campbell, GM Safety Research and Development Laboratory, July 1974, SAE 740565.

This is the paper upon which some curent computer simulation programs are based. Mr. Campbell assumes that the force required to cause crush is a linear function of the crush depth, that is, "F = mc + b," where "m" and "b" are constants associated with the particular vehicle, and "c" is the crush depth. A discussion of this is provided in Chapter 10.

Automobile Side Impact Collisions - Series II by Severy, Mathewson and Siegel, University of California at Los Angeles, SAE SP-232.

Excellent monograph on side collisions, especially their effects on vehicle occupants.

D. Regular Sources of Information

Consumer Reports

Various issues during each year discuss performance factors of popular cars and vehicles. Information regarding acceleration, braking distances, curb weight, and so on is listed.

Insurance Institute for Highway Safety/Highway Loss Data Institute.

These two organizations are essentially the same, and are supported by the insurance industry. They collect data on automotive safety topics, e.g. crash rates, relative crashworthiness, effects of regulations, anti-lock brake devices, and so on. They publish several papers and brochures each year on their findings, and related trends.

E. Sample Vehicular Accident Reports

The following six pages show three blank vehicular accident reports of the type used in the state of Kansas. They are similar to forms used in many other states.

The forms are included so that the reader can examine how such forms are organized, what information they contain, and the type of codes and shorthand notes that are used to fill in the blanks. From such standardized reports, accident statistics are compiled.

Typically, the information contained in such a report will form a major part of the information available to the engineer to assess the accident.

STATE OF KANSAS
MOTOR VEHICLE ACCIDENT REPORT

(Check One Box Only) ☐ Fatal ☐ Injury ☐ Private Property ☐ Property Damage Only **OVER** $500 ☐ Property Damage Only **UNDER** $500

Hit & run Accident?

Milepost	**COUNTY**	**ON** Highway/Road/Street & Speed Limit:	**CITY/TOWN**	Photos By	Local Case Number	Sheet of /
Distance Ft/Mi & Dir		☐ **FROM** ☐ **AT** Highway/Road/St & Speed Limit:	Investigating **DEPT.**	Investigating **OFFICER**		Reviewed By

COLLISION DIAGRAM (Show Unit Movements, Roads, North Arrow) (N)

DESCRIBE PRE-CRASH MOVEMENT or **ACTION** and **DIRECTION** of **VEHICLES** and **PEDESTRIANS/PEDALCYCLISTS** By Number

DATE of **ACCIDENT**

TIME Occurred & **DAY**

TIME Notified/Arr. & **DAY**

Object Damaged & Nature of Damage (Show Location in Diagram)

Name and Address of Object Owner

(State Use Only)

ON Road	Cntl Sec	Sec. Milepost	Int #	AT Road	Distance	Unit	Dir	Latitude	Longitude

County	City Code	Agency Code N	Distance M	Reference Road 1 +	E Distance M	Reference Road 2	Coder	F A Code

Unit	☐ Driver ☐ Ped **NAME** (Last, First and Initial)	Phone ☐ Work ☐ Home	Color	**YEAR & MAKE** of Vehicle	**MODEL** and **BODY STYLE**

Driver/Ped **ADDRESS** (Number, Street, City, State, Zip Code)	**STATE**	**LICENSE PLATE #**	**YEAR**	Removed By:

DRIVER'S LICENSE STATE and **NUMBER** St. | No.	**CDL?**	**DATE** of **BIRTH**	**SEX**	**VEHICLE IDENTIFICATION NUMBER**	Odometer

Registered **OWNER FULL NAME** ("Same" if Driver)	Phone ☐ Work ☐ Home	**TOTAL** number of vehicle occupants including driver	Fire?	Insurance Company

OWNER Address ("Same" if Driver)	Special Data Area	Policy Number

Special Conditions (if any) for unit above: ☐ 01 Hit & Run ☐ 02 Non-Contact ☐ 03 Stolen ☐ 04 Parked ☐ 05 Police Pursuit ☐ 06 Driverless

Unit	☐ Driver ☐ Ped **NAME** (Last, First and Initial)	Phone ☐ Work ☐ Home	Color	**YEAR & MAKE** of Vehicle	**MODEL** and **BODY STYLE**

Driver/Ped **ADDRESS** (Number, Street, City, State, Zip Code)	**STATE**	**LICENSE PLATE #**	**YEAR**	Removed By:

DRIVER'S LICENSE STATE and **NUMBER** St. | No.	**CDL?**	**DATE of BIRTH**	**SEX**	**VEHICLE IDENTIFICATION NUMBER**	Odometer

Registered **OWNER FULL NAME** ("Same" if Driver)	Phone ☐ Work ☐ Home	**TOTAL** number of vehicle occupants including driver	Fire?	Insurance Company

OWNER Address ("Same" if Driver)	Special Data Area	Policy Number

Special Conditions (if any) for unit above: ☐ 01 Hit & Run ☐ 02 Non-Contact ☐ 03 Stolen ☐ 04 Parked ☐ 05 Police Pursuit ☐ 06 Driverless

Record **ALL** injured & uninjured vehicle occupants and pedestrians. For trains, record crew in control and injured passengers. See Code List.

TRAF UNIT	**SEAT/ TYPE**	Use Codes for Vehicle Occupants, Pedestrians & Cyclists. List Witnesses as W1, W2, etc. Last **NAME** First Name Initial	**ADDRESS** (#, Street, City, State, Zip.)	**SEX**	**AGE**	**S.E. USE**	**Child Youth**	**EJCT TRAP**	**INJ. SEV.**	**EMS UNIT**

EMS Unit	**INJURED TAKEN** By:	**INJURED TAKEN** To:	EMS Unit	**INJURED TAKEN** By:	**INJURED TAKEN** To:	EMS Unit	**INJURED TAKEN** By:	**INJURED TAKEN** To:
A			B			C		

SPECIAL DATA *(State use Only)* | Microfilm Reference # | State Accident Case #

Dr/Pd #	Violation Charged	Citation No.	Dr/Pd #	Violation Charged	Citation No.	Dr/Pd #	Violation Charged	Citation No.

OFFICER'S OPINIONS OF APPARENT CONTRIBUTING CIRCUMSTANCES (Factor Type-Unit Number/Specific Factor) Enter in order all codes that apply.

LIGHT
01 Daylight
02 Dawn 99 Unknown
03 Dusk
04 Dark-street lights on
05 Dark-no street lights

WEATHER/ATMOSPHERE
00 No adverse condition
01 Rain
02 Sleet 14 Rain & Fog
03 Snow 16 Rain & Winds
04 Fog 24 Sleet & Fog
05 Smoke 36 Snow & Winds
06 Severe winds
07 Blowing dust/sand/etc.
88 Other ______

SURFACE TYPE (ON / AT)
01 Concrete
02 Blacktop
03 Gravel
04 Dirt
88 Other ______

SURFACE CONDITION (ON / AT)
01 Dry
02 Wet
03 Snow or slush
04 Ice
05 Mud, dirt or sand
06 Debris (Oil, etc.)
88 Other ______

TRAFFIC CONTROLS
O/A (On/At) Road
Type Present
OK/NF(OK/Non-Functional)
00 None
01 Officer, flagger
02 Traffic signal
03 Stop sign
04 Flasher
05 Yield sign
06 RR Gates or signal
07 RR Crossing signs
08 No passing zone
09 Center/edge lines
88 Other ______

ROAD CHARACTER (ON / AT)
01 Straight and level
02 Straight on grade
03 Straight at hillcrest
04 Curved and level
05 Curved on grade
06 Curved at hillcrest
88 Other ______
99 Unknown

CONST./MAINT. ZONE (ON / AT)
00 None apply
01 Construction Zone
02 Maintenance Zone
03 Utility Zone

ACCIDENT CLASS
00 Other Non-collision
01 Overturned
COLLISION WITH:
02 Pedestrian
03 Other Motor Vehicle*
04 Parked Motor Vehicle
05 Railway Train
06 Pedalcycle (Bicycle)
07 Animal (Specify) ______
08 Fixed Object **
09 Other Object ______

ACCIDENT LOCATION
ON ROADWAY:
11 Non-intersection
12 Intersection
13 Intersection-related
14 Parking/driveway access
15 Interchange area
16 On Crossover
OFF ROADWAY:
21 Roadside (incl. shoulder)
22 Median
88 Other ______
99 Unknown

* COLLISION WITH OTHER MOTOR VEH.
01 Head on
02 Rear end
03 Angle
04 Sideswipe-opposing
05 Sideswipe-overtaking
88 Other ______
99 Unknown

** FIXED OBJECT TYPE
01 Bridge Structure
02 Bridge Rail
03 Crash Cushion (Barrels)
04 Divider/Median Barrier
05 Overhead Sign Support
06 Utility Pole
07 Other Post/Pole
08 Building 14 Hydrant
09 Guardrail 15 Barricade
10 Sign Post 16 Mailbox
11 Culvert 17 Ditch
12 Curb 18 Embankment
13 Fence 19 Wall
20 Tree
88 Other ______
99 Unknown

ROAD SPECIAL FEATURES
Identify all that apply
00 None 04 Railroad crossing
01 Bridge 05 Interchange
02 Bridge overhead 06 Ramp
03 Railroad bridge 88 Other ______

Enter any visible identifier; refer by code.
Code: ______ Ident: ______

VEHICLE MANEUVER BEFORE CRASH
01 Straight/following road
02 Left turn
03 Right turn
04 U turn
05 Overtaking (passing)
06 Changing lanes
07 Avoiding maneuver
08 Merging
09 Parking
10 Backing
11 Stopped awaiting turn
12 Stopped in traffic
13 Parked
88 Other ______
99 Unknown

VEHICLE DAMAGE
00 None/None known
01 Damage (Minor)
02 Functional
03 Disabling
04 Destroyed
88 Other ______
99 Unknown

DAMAGE LOCATION AREA--Vehicle 1

FRONT

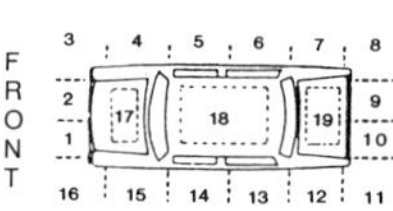

☐ Top ☐ Windshld ☐ Windows
☐ Under ☐ Overturn ☐ Towed

DAMAGE LOCATION AREA--Vehicle 2

FRONT

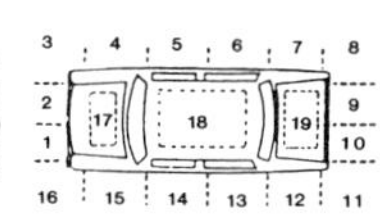

☐ Top ☐ Windshld ☐ Windows
☐ Under ☐ Overturn ☐ Towed

SCHOOL BUS CAPACITY

MOTORCYCLE CCs

VEHICLE BODY TYPE
01 Automobile 10 *Single truck over 4-tires**
02 Motorcycle 11 *Truck and trailer(s)**
03 Motorsctr/Moped 12 *Tractor-trailer(s)**
04 Van 13 *Cross country bus**
05 Pickup truck 14 *School bus**
06 Single truck 4-tires 15 *Transit (city) bus**
07 Camper/RV * Truck Report required
08 Farm Equipment 88 Other ______
09 All Terrain Vehicle

TRAILER?
1 ☐Present ☐Damaged
2 ☐Present ☐Damaged

PED/CYCLIST LOCATION BEFORE IMPACT --
IN INTERSECTION WITH CROSSWALK/BIKEWAY:
01 In crosswalk/bikeway
02 Not in crosswalk/bikeway
03 In intersection without crosswalk/bikeway
NOT IN INTERSECTION BUT ELSEWHERE IN ROADWAY:
11 In available crosswalk or bikeway
12 Not in available crosswalk or bikeway
13 In area without crosswalk or bikeway
25 NOT IN ROADWAY

PED/CYCLIST ACTION
01 Entering/crossing road
02 Walking/riding on road
03 Approaching, leaving, or working on vehicle
04 Work (not on vehicle)
05 Playing or standing
06 Approaching/leaving bus
88 Other ______
99 Unknown

PED/CYC OBEDIENCE TO PED TRAFFIC SIGNAL
00 No pedestrian signal
01 Obeyed pedestrian signal
02 Disobeyed ped. signal
03 Ped. signal malfunction
99 Unknown

DR. LIC. COMPLY
Code each Driver
00 Not licensed
01 Valid license
02 Invalid license

RESTRICT. COMPLY
00 No restrictions
01 Complied with
02 Did not comply
99 Unknown

SUBSTANCE USE & CONTRIBUTION
(2-Character Code)
SUBSTANCE CONTRIBUTION
A - Alcohol P - Present
D - Illegal Drug C - Contributed
M - Medication to Accident

DRIVER/PED IMPAIRMENT TEST
TR Alcohol/Drug Test Refused
PT Positive Preliminary Test
RP Evidentiary Test Given and Results are Pending
<--Alcohol Test Results

TRUCK-FATALITY REPORT

COUNTY	ON Highway/Road/Street	CITY/TOWN	DATE of Accident	☐ Supplement ON TRUCK ACCIDENT ☐ Supplement ON FATAL ACCIDENT ☐ Supplement ON TRUCK & FATAL	Sheet of /
FOR STATE USE ONLY		INVESTIGATING DEPT.	TIME Occurred & DAY	Invest. OFFICER	Local Case Number

The State requires this additional report for accidents involving trucks*, buses*, or fatalities**. Local policies may also require it.

TRUCK DATA

*A Truck Report is required if **TRUCK** has at least dual axles and 6 tires or if **BUS** capacity is 15 or more seats. MVAR Unit #

NAME OF CARRIER (CORPORATE BUSINESS NAME)

KANSAS PERMITS (Issuer and Permit Number)
1.
2.
3.

CARRIER ADDRESS CITY STATE ZIP CODE

U.S. GOVERNMENT PERMITS (Issuer and Number)

U.S.DOT: ICC MC:

SOURCE OF NAME (Enter only one)
01 Side of Vehicle 03 Driver
02 Shipping Papers or Manifest

TRAFFICWAY TYPE
00 Undivided
01 One Way Roadway
02 Divided Highway, Median Strip without Barrier
03 Divided Highway, Median Strip with Barrier

ACCESS CONTROL
00 No Control (Unlimited Access) 88 Other
01 Full Control (Entry/Exit only by ramp)

VEHICLE DRIVEN Use only 1 of next 3 lines	Cab	Number of Axles	Number of Wheels	TRUCK BODY TYPE CODE (Describe if "Other")	Length of Unit (Feet)	Width of Unit (Inches)	Gross Vehicle Weight	Cargo Type Code	HAZARDOUS MATERIALS DATA: Placard Color	Placard Number	Spill or Release?
Single Truck (01)											
Tractor (02)				02							
Bus (03)				Number of Seats in Bus (Capacity)							
TRAILERS (if any) Use as many lines as necessary to include all trailers in use at the time of the accident											
Semi-trailer (04)											
Full Trailer (05)											
2nd Trailer (06)											
3rd Trailer (07)											

CODES

CAB CODE (for Single Truck or Tractor): 01 = Cab **Behind** Engine 02 = Cab **Over** Engine

UNIVERSAL CODES: 88 = Other 99 = Unknown

TRUCK BODY TYPE
01 Van/Enclosed Box
02 Truck-Tractor
03 Cargo Tank
04 Flatbed
05 Dump
06 Concrete Mixer
07 Auto Transporter
08 Garbage/Refuse
15-50 # of Bus Seats

CARGO TYPE
00 Empty
01 Driveaway/towaway
02 Explosives
03 Farm & Other Animals
04 Farm Products
05 Gases (Bulk)
06 General Freight (Pkgs)
07 Heavy Machinery/Objects
08 Household Goods
09 Liquids (Bulk)
10 Logs/Poles/Lumber
11 Metal (Coils/Sheets/Etc.)
12 Mobile Home
13 Motor Vehicles
14 Refrig. Foods
15 Solids (Bulk)
88 Other
99 Unknown

SEQUENCE OF EVENTS (List 4 if 4 apply.)
1
2
3
4
00 Ran Off Road
11 Jackknife
12 Overturn
13 Downhill runaway
14 Cargo loss/shift
15 Explosion
16 Fire
17 Separation of units

Collision with:
21 Pedestrian
22 Motor Veh. in trans.
23 Parked Motor Vehicle
24 Train
25 Pedalcycle
26 Animal
27 Fixed Object
28 Other Object
88 OTHER EVENT

FATALITY DATA

**Copies of all narratives and diagrams made by investigating officers are also required for all accidents involving fatalities.

VEHICLE 1 DAMAGE

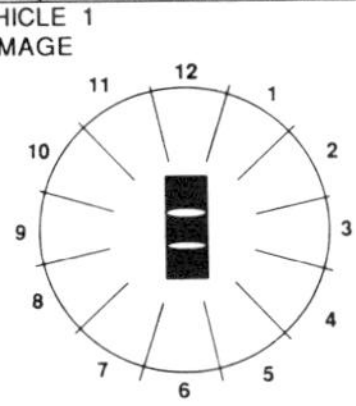

☐ Undercarriage
☐ No Damage
Estimated Speed, MPH

TIME EMS NOTIFIED

TIME EMS ARRIVED

TIME EMS ARRIVED AT HOSPITAL

SPECIAL JURISDICTION
00 Not Special
01 Nat. Park Serv.
02 Military
03 Indian Reserv.
04 College/University Campus
05 Other Federal Properties
88 Other
99 Unknown

EXTRICATION WAS REQUIRED FOR THE FOLLOWING PERSONS

IMPACT POINTS Show both initial impact point(s) by arrow(s) and principal impact point(s) by arrow(s), and label each arrow "**I**" or "**P**" to indicate the proper type.

VEHICLE 2 DAMAGE

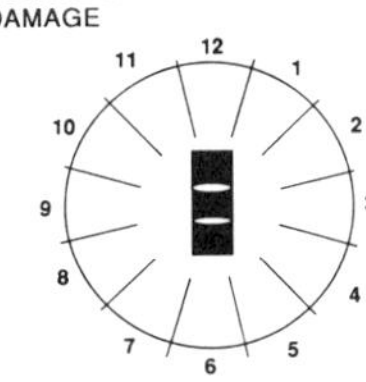

☐ Undercarriage
☐ No Damage
Estimated Speed, MPH

USE THE **INVESTIGATIVE REPORT FORM** AS NEEDED FOR NARRATIVE EXPLANATION AND DIAGRAMMING

APPARENT CONTRIBUTING CIRCUMSTANCES CODES
(LIST IN ORDER OF SIGNIFICANCE)

D (n)-**DRIVER** (1,2,etc.)

01 Under influence of drugs
02 Under influence of alcohol
03 Failed to yield right of way
04 Disregarded traffic signs, signals, or road markings
05 Exceeded posted speed limit
06 Too fast for conditions
07 Made improper turn
08 Wrong side or wrong way
09 Followed too closely
10 Improper lane change
11 Improper backing
12 Improper passing
13 Improper or no signal
14 Improper parking
15 Fell asleep, fainted, ill, etc.
16 Failed to give full time and attention
17 Did not comply with license restrictions
18 Interference/Obstruction by passenger

P (n)-**PEDESTRIAN/CYCLIST** (1,2,etc.)

01 Under influence of drugs
02 Under influence of alcohol
03 Failed to yield right of way
04 Disregarded traffic control
05 Illegally in roadway
06 Bicycle violation
07 Clothing not visible

V (n)-**VEHICLE** (1,2,etc.)

01 Brakes
02 Tires
03 Exhaust
04 Lights
05 Windows/windshield
06 Wheel(s)
07 Trailer coupling
08 Cargo
09 Unattended (in motion)

O/A(On/At) **R**(Road)

01 Wet
02 Icy or slushy
03 Debris or obstruction
04 Ruts, holes, bumps
05 Road under construction or maintenance
06 Traffic control device inoperative
07 Shoulders: low/soft/high

E - ENVIRONMENT

01 Smog, smoke
02 Sleet, hail, freezing rain
03 Blowing sand, soil, dirt
04 Severe crosswinds
05 Rain, snow
06 Animal
07 Vision obstruction (including building, sun, trees, etc.)

Contributing Circumstances Example	**OFFICER'S OPINIONS . . .**				Interpretation:
	D1	*07*	*OR*	*02*	Dr 1: Impr. Turn (07) On Rd: Icy/Slush (02)

TRUCKS NOT REQUIRING REPORT

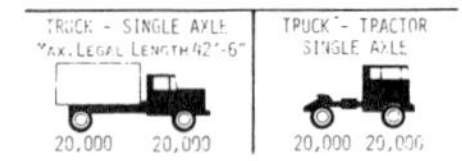

TRUCKS REQUIRING REPORT

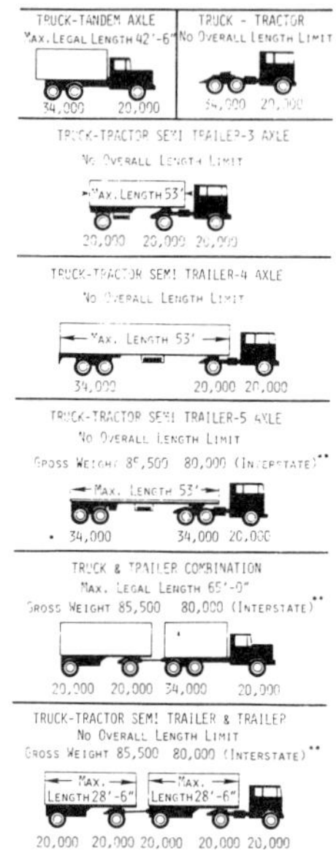

SEAT CODES FOR VEHICLE OCCUPANTS

SEAT (Location)
01 **DRIVER** (or motorcycle operator)
02 Center front
03 Right front
04 Left rear (or motorcycle passenger)
05 Center rear
06 Right rear
07 Other seat position **IN** vehicle (or 2nd motorcycle passenger)
08 Any position **ON** or **OUTSIDE** of vehicle including in truck bed
09 Unknown location **IN** or **ON** vehicle
11 Extra person in driver's seat or on driver's lap
12 - 17 Extra person in seat or on lap of seated passenger

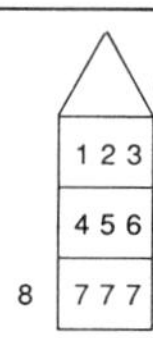

CODES FOR TRAIN OCCUPANTS

31 Train Crew (List all in control whether injured or not)
32 Train Passenger (List only if injured)

PEDESTRIAN/CYCLIST TYPE CODES

21 Pedestrian
22 Pedalcyclist
23 Rider of animal
24 In animal-drawn vehicle
25 In vehicle **NOT IN TRANSPORT**
26 Machine operator/rider
88 Other
99 Unknown

SAFETY EQUIPMENT USE

N None used	**P** Passive system
S Shoulder/lap	**L** Lapbelt only
H MC Helmet	**C** Ch/youth restr.
E MC Eye protect.	**U** Unknown

CHILD/YOUTH RESTR. *MISUSE*

OK Use **OK** (Not misused)
UN Unused *But Present in Vehicle*
FW Device Facing Wrong Way
NA Not Anchored Properly
NS Anchor Not Secure
NT Not Tethered Properly
SL Strap/tether Loose
ST Size/Type Improper for Child

INJURY SEVERITY

N Not injured/not known
P Possible injury
I Injury, not incapacitating
D Disabled (incapacitating)
F Fatal Injury

EJECTED/TRAPPED **N**-No **E**-Ejected **T**-Trapped

SEX **M**-Male **F**-Female

INVESTIGATIVE REPORT

COUNTY	ON Highway/Road/Street	CITY/TOWN	DATE of Accident	☐ Narrative & Diagram on FATAL ACCIDENT (Required by State) ☐ Investigative Report		Sheet of /
FOR STATE USE ONLY		INVESTIGATING DEPT.	TIME Occurred & DAY	Invest. OFFICER	Local Case Number	

DIAGRAM WHAT HAPPENED	Draw scene as observed. Refer to vehicles, drivers and pedestrians by numbers assigned in this report.

SHOW (1) Outline of street and access points and identify specifically by number.
(2) Paths of units prior to and after impact, skidmarks, and point of impact (POI).
(3) Location of signs, traffic controls, and reference points.
(4) Location of other property hit/damaged (trees, signs, etc.).
(5) Special features at location (bridge, overpass, culvert, railroad crossing, etc.).
(6) Location of temporary highway conditions.
(7) All measurements to locate the accident relative to specific, fixed, uniquely identifiable and locatable point.

Indicate NORTH by Arrow

Index